Adequate Wisdom

Adequate Wisdom

Essays on the Nature of Existence

A Layman's Observations
of Life & the Cosmos

Ronald P. Smolin

BAINBRIDGEBOOKS
Philadelphia 2012

Published by BainBridgeBooks,
an imprint of
Trans-Atlantic Publications Inc.
Philadelphia PA USA

www.adequatewisdom.org

ISBN 9781891696305

Library of Congress Cataloging-In-Publication Data

Smolin, Ronald P.
  Adequate wisdom : essays on the nature of existence : a
  layman's observations of life & the cosmos / Ronald P.
  Smolin
  p. cm.
  Summary: "Provides a broad overview of the structures,
  events and ideas in the world. Includes sections on
  physical and biological existence, God and religion,
  and the human condition"—Provided by the publisher.
  Includes bibliographical references (p.
  ISBN 978-1-891696-30-5 (alk. paper)
1. Life. 2. Philosophy. I. Title
  BD431.S593 2012
     191—dc23
                    2011037243

Manufactured in the United States of America

First edition

To all members of our species

Preface

This work contains the observations by one member of our species who has forever stood in awe of his own existence and that of the entire universe, trying to understand why things are the way they are.

One way to examine existence is to look at the structures, events and ideas in the world and comment on them; including their effects upon the day-today activities of humankind.

Of course it is virtually impossible to codify existence, but nonetheless a humble attempt is made in this work, because curiosity urges one to fashion an integration of life and the cosmos as best as possible.

The goal is the creation of an adequate wisdom (a workable wisdom) to help us formulate policies and judgments that benefit not only our species but the entire planet as well.

Overview of Contents

PART ONE:
THE ESSENTIALS

PART TWO:
QUESTIONS & IDEAS

PART THREE:
PHYSICAL EXISTENCE

PART FOUR:
BIOLOGICAL & HUMAN EXISTENCE

PART FIVE:
TRENDS & OTHER MATTERS

PART SIX:
GOD & RELIGION

PART SEVEN:
HUMANITY

PART EIGHT:
FINAL THOUGHTS

Glossary

Bibliography

Table of Contents

PART ONE: THE ESSENTIALS

PART TWO: QUESTIONS AND IDEAS

PART THREE: PHYSICAL EXISTENCE

PART FOUR: BIOLOGICAL & HUMAN EXISTENCE

PART FIVE: TRENDS & OTHER MATTERS

PART SIX: GOD & RELIGION

PART SEVEN: HUMANITY

PART EIGHT: FINAL THOUGHTS

GLOSSARY

BIBLIOGRAPHY

Introduction

This work attempts to present an overview of existence that may assist general readers who want to know more about the world and their place in it. We can begin to piece together the varied components of existence, creating a clearer understanding of how the world works and then proposing guidelines to help us make wise decisions and lead meaningful lives.

This overview presents key tools that should enable the reader to gain a broader understanding of the world and thus establish the beginnings of an adequate wisdom, a phrase that is being used in this work to denote a collection of knowledge that assists us in creating commonsensical and rational policies and beliefs that contribute to the flourishment of the biosphere, hydrosphere and atmosphere.

The tools serve as guideposts which allow the reader to organize the enormous number of structures, forces and thoughts that surround us. By utilizing the tools, we should be able to obtain a big picture of the world which hopefully will produce positive actions and thoughtful moral judgments.

Even though the approach at times will concentrate on specific entities, we must never keep our eye off the interrelatedness of existence and the continuous interplay between the fixity and fluidity of the universe.

These essays are brief introductions to observations about the world and should therefore be read as foundations for study and criticism. Because of the nature of the subject, some essays will contain speculative ideas.

ADEQUATE WISDOM

PART ONE:

THE
ESSENTIALS

The essays in Part One focus on the approaches we can take to create an overview of the various parts of existence, including forms, processes and ideas; the modes of determinism, chance, design, free will and synergy; and the division of the world into the physical, biological, personal and social realms. There are twelve variables of existence that are shown to create an interrelationship that accounts for our experiences and knowledge. The influence of recurring forms, processes and ideas is discussed. Also, there are essays discussing the problems we face in trying to gain a comprehensive view of life and the cosmos, as well as a discussion of the status of the human condition.

The grandeur and mystery of existence

How can any of us truly understand the wonder of the universe, of the spectacular presence of trillions of stars, billions of galaxies and some nine million living species on our planet? How do we begin to understand the dynamics of human relationships and interactions?

Our minds are besieged with an endless stream of structures and forces and ideas, and we grasp for any semblance of clarity and understanding of the world engulfing us.

To exist is remarkable. To think is remarkable. To experience a lifetime of emotions, personal relationships and accomplishments is remarkable.

We scurry about doing our business, setting and reaching goals, hopefully caring for one another and enjoying all that life offers. Some of us stop to think occasionally about the world and our place in it and speculate about why things are the way they are. We ask such questions as: What's it all about? Am I doing the right thing? What *is* the right thing? Am I following what everyone does or am I setting out on paths different from others? What influences me? What influences others? What influences civilization? What influences the cosmos?

To ask such questions sets the framework for the construction of an adequate wisdom that incorporates the rational and commonsensical view of existence. Such wisdom seeks to establish a broad overview of life and the universe.

At first glance, existence appears to be an impenetrable and befuddling maze. But the cognitive power of the brain can chip away and find some openings through which we may discern certain patterns that partially explain the world and our very lives, leading us towards the evolution of a humanistic civilization.

Forms, processes and ideas

To grasp the enormity of existence, we need to establish some means by which we can look at the big picture and see how things may be related to one another. The following tools attempt to accomplish that task.

Forms comprise matter, structures and wholes; such as atoms, stars, cells, planets, living things and social institutions.

Processes comprise energy, forces, actions and events; such as gravity, evolution, consciousness, policy making, social intercourse and electromagnetism.

Ideas represent beliefs, values and common sense. Ideas result from the mental constructs resulting from the observation of forms and processes; such as the ideas created by mathematics, physics, chemistry, biology, religion, cosmology, philosophy, sociology and psychology.

Ideas also allow for metaphysical interpretations of existence so that concepts like God, mysticism, vitalism, purpose and other subjects may be considered.

These tools enable us to view the world in the physical, biological, personal and social frames of reference which form the bases of an adequate wisdom of existence.

It is the contention of this work that forms, processes and ideas allow us partially to understand how the world works and how everything we experience is governed by their interactions.

THE ESSENTIAL OVERVIEW OF EXISTENCE

To help us view existence, we have established three tools that represent the world as we know it. Forms are structures like stars, cells, and human beings. Processes are the energy, actions and events in the world, such as evolution, consciousness and electromagnetism. Ideas result from our observations of forms and processes and allow us to create opinions, values, scientific and philosophical laws and theories. These three tools provide the overview for the development of adequate wisdom.

They are inexorably linked together.

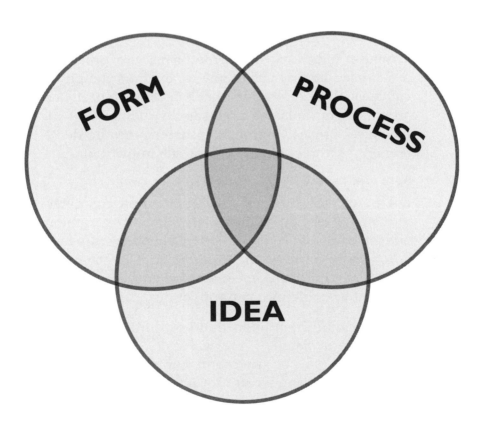

Modes of existence

The modes of existence impact the forms, processes and ideas. The study of them will allow us to make significant inroads toward our quest for adequate wisdom.

Design indicates a plan for the construction of a form and/or process, such as a watch, a computer program, a legislative action, a pharmaceutical drug, a screenplay, a sporting event, an airplane or perhaps a universe. Design manifests itself in the creation of artifacts, as well as the laws, policies and regulations that greatly influence the behavior of human beings.

Determinism indicates that some forms and processes are caused by prior phenomena, such as the physical laws that determine the behavior of particles, atoms, stars, galaxies and solar systems; and the biological processes that help determine the development of whole species and single individuals.

Chance represents those processes and events that are not predictable and exist without apparent determinism or design. Chance indicates the random crisscrossing of events or even a single event whose outcome we could not have predicted. Chance represents uncertainty. (Some chance events may be part of a pattern which we cannot yet discern.)

Free will permits the human mind to think, act, create and make decisions. Free will can initiate design, so the two may be interchangeable at times. Free will does not mean we are totally free of constraints placed upon the individual mind and body, as we are influenced by all other forms, processes and ideas. But free will provides the means to pursue pure thought and volition.

Synergy represents forms (wholes) and processes (events) and their tendency to combine into more complex forms and processes. Each structure has its own identity, its own encapsulation, its own functions. Stable atoms are unique forms. So too are the various types of cells in living beings. So too is each star, planet, galaxy, person, family, corporation and nation.

These individual forms and processes coalesce, so that when individual structures are combined, they form a new, unexpected structure, greater than and different from each one if it had stood on its own. Forms and processes often sort themselves into hierarchical arrangements.

Understanding synergy and its role in the universe will shed light on the evolution of the cosmos, as well as the evolution and behavior of living things.

THE INTERPLAY AMONG TOOLS

It is most important to understand that while the tools can be studied individually, they actually are interconnected and, when viewed as an entirety, enable us to create an adequate wisdom of existence. *Process and form depend upon each other and therefore we cannot have one without the other.*

The use of the tools does not mean that we can establish neat packages to explain all of existence. On the contrary, the cosmic world and the living world are spectacularly complex. It's one thing to look at an overview of existence; it's another thing to think that we can truly grasp it all. But surely, as inquisitive beings, it is our obligation to make some sense of this remarkable universe.

THE MODES OF EXISTENCE

These are the five modes by which form, process and idea are affected. Synergy results from the combination of two or more forms and processes. Determinism represents that which causes other things to happen. Design is the plan for the creation of forms and processes and ideas. Chance represents uncertainty. Free will enables the human mind to think and act.

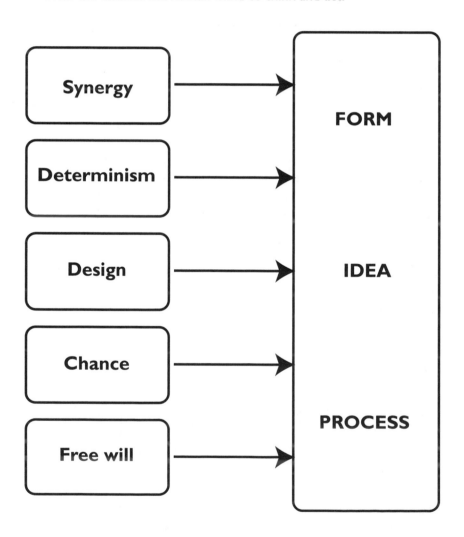

The four divisions of existence

To help sort through the varieties of forms, processes and ideas, we utilize the following divisions.

Physical Existence This includes the inorganic world of particles, atoms, stars, planets and galaxies, and all the physical forces that helped to bring about the evolution of matter. This division includes the Sun, planet Earth and other celestial bodies in the solar system. It includes the Earth's solid core, molten outer core, the mantle, the plates and crust upon which rest the continents, as well as all water systems and atmospheric layers. It also includes compounds and structures created by humankind, such as pharmaceuticals, electronic equipment and skyscrapers.

Biological Existence This division contains the biosphere and all living things, below, on and above the Earth. The biosphere owes its existence to the physical world and interacts with it in many ways.

Personal Existence (Self) Each individual human being exhibits personality, emotion, volition, character and talent. Each person interrelates with the biosphere, the social sphere and the physical world.

Social Existence Here we have all different combinations of humans in various bonds, alliances and social structures. The social sphere interacts with all other divisions.

The variables of existence

1. Physical forms
2. Physical processes
3. Physical ideas

4. Biological forms
5. Biological processes
6. Biological ideas

7. Personal forms
8. Personal processes
9. Personal ideas

10. Social forms
11. Social processes
12. Social ideas

These 12 categories comprise the basis of an adequate wisdom. They all interact and they all must be considered when formulating judgments, thoughts, decisions, ethics and ultimately an approach to truth. They also exhibit hierarchies as well as synergistic assemblages.

We, as individual human forms, are beholden to the variables that precede us, such as the physical, biological and social forms and processes; and yet we stand out as synergistic wholes that transcend these determinants.

Excluding any of the 12 variables from our policies and decisions attenuates our approach to adequate wisdom.

The Variables of Existence

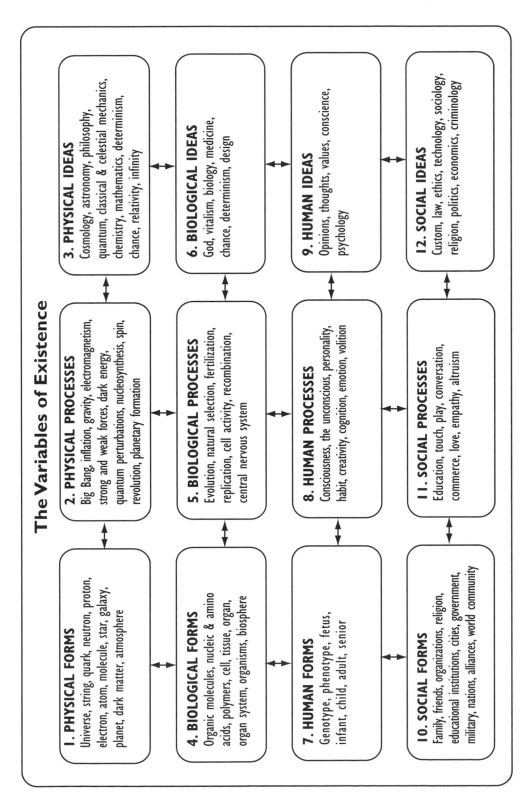

1. PHYSICAL FORMS
Universe, string, quark, neutron, proton, electron, atom, molecule, star, galaxy, planet, dark matter, atmosphere

2. PHYSICAL PROCESSES
Big Bang, inflation, gravity, electromagnetism, strong and weak forces, dark energy, quantum perturbations, nucleosynthesis, spin, revolution, planetary formation

3. PHYSICAL IDEAS
Cosmology, astronomy, philosophy, quantum, classical & celestial mechanics, chemistry, mathematics, determinism, chance, relativity, infinity

4. BIOLOGICAL FORMS
Organic molecules, nucleic & amino acids, polymers, cell, tissue, organ, organ system, organisms, biosphere

5. BIOLOGICAL PROCESSES
Evolution, natural selection, fertilization, replication, cell activity, recombination, central nervous system

6. BIOLOGICAL IDEAS
God, vitalism, biology, medicine, chance, determinism, design

7. HUMAN FORMS
Genotype, phenotype, fetus, infant, child, adult, senior

8. HUMAN PROCESSES
Consciousness, the unconscious, personality, habit, creativity, cognition, emotion, volition

9. HUMAN IDEAS
Opinions, thoughts, values, conscience, psychology

10. SOCIAL FORMS
Family, friends, organizations, religion, educational institutions, cities, government, military, nations, alliances, world community

11. SOCIAL PROCESSES
Education, touch, play, conversation, commerce, love, empathy, altruism

12. SOCIAL IDEAS
Custom, law, ethics, technology, sociology, religion, politics, economics, criminology

The guiding principles of adequate wisdom

There are four guiding standards upon which adequate wisdom relies and which help us achieve a balanced and worthwhile lifetime. They are stated here with a brief introduction.

Pleasure is the enjoyment of the bounty of nature, personal relationships, creativity, sexuality, humor, the arts, conversation, gustation and the rest of life's treasures and pleasures.

Responsibility requires us to establish guidelines that keep us in check with the world around us. Responsible people do not run amok; they exhibit self-control. They balance their lives so that most actions are moderate and belief systems are open to new ideas.

Strength is necessary to protect oneself and one's friends, family and world community. It also allows us to act with tenacity in all our endeavors and provides the courage of our convictions if they are based upon commonsense and rationality.

Compassion requires sensitivity to all people and all living things. Compassion results from understanding the trials and tribulations of other people. It is based upon the adequate wisdom we have accumulated. Compassion is the greatest virtue as it fosters care, concern, protection and love for all living creatures.

BALANCING THE PRINCIPLES OF ADEQUATE WISDOM

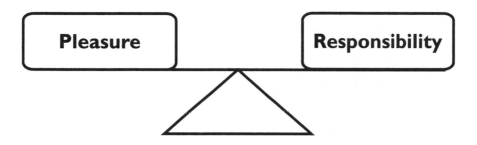

To live without the pleasure obtained from the joys of creativity, love, humor, laughter, sexuality, sport and entertainment bodes poorly for most individual humans. To live without the application of responsibility leads to a reckless existence, without planning, organizing and self-control.

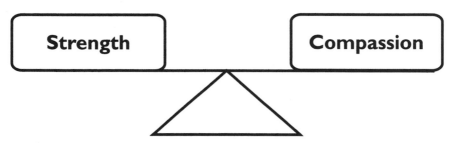

To live without attaining the strength to protect oneself, one's friends, family and world community from the ravages of chance and determinism leaves one vulnerable to those who prey upon others. To live without compassion and love is the gravest omission of all, since one cannot appreciate the need to take care of our species and to foster the flourishment of all species and our planet.

A life of wisdom must incorporate all four components harmoniously balanced.

Steps to adequate wisdom

1. We observe the forms and processes in the world.

2. We analyze these forms and processes to create ideas about them.

3. We begin to sort these forms, processes and ideas and organize them into physical, biological, personal and social categories.

4. We examine the modes of existence, including chance, determinism, design, free will and synergy.

5. We make assessments and assumptions to form the basis of adequate wisdom.

6. We apply adequate wisdom to make beneficial policies and judgments for the preservation and flourishment of the biosphere and the planet as a whole, based upon the standards of pleasure and responsibility and strength and compassion.

Adequate wisdom requires one to look at the big picture and to reflect upon the key ideas of existence. It requires one to understand the contextual relationships among major structures and processes and view everything from different frames of reference.

Adequate wisdom should allow us to form a basis of morality and to use our minds to make judgments and decisions that align with the reality we perceive and the one which we try to shape.

The more we understand the interplay and dynamical relationships of structures, events and ideas, the more we approach an adequate wisdom.

The scales of existence

GRAND SCALE OF EXISTENCE

The multiverse (?)

Our universe

Dark energy

Dark matter

Super galactic clusters

Clusters of galaxies

Galaxies

LARGE SCALE OF EXISTENCE

Open clusters of stars

Closed clusters of stars

Stars & Nebulae

Black holes

SOLAR SYSTEM

Local star

Planets

Moons

Other celestial bodies

PLANETARY SCALE

Core

Outer core

Mantle

Crust

Hydrosphere

Atmosphere

Magnetosphere

BIOLOGICAL EXISTENCE

All living forms in the biosphere.

SOCIAL EXISTENCE

Families

Friends

Groups

Associations

Institutions

Nations

National collectivities

Civilizations

PERSONAL EXISTENCE

Each self

Consciousness

Unconscious

Personality

Emotion and Habit

Thought

Action

MICROSCOPIC & SUBMICROSCOPIC EXISTENCE

Strings

Particles

Atoms

Waves

Molecules

Living cells

The major forms

Here are the major forms comprising existence. They are presented only as a listing to provide one with an appreciation of the depth of the forms. Some will be mentioned in later essays or the glossary.

String	Polymer	Bacteria
Higgs boson	Nucleic acid	Fungus
Quark	Amino acid	Multicellular life
Gluon	DNA	Tissue
Graviton	RNA	Neuron
Proton	Gene	Brain
Neutron	Allele	Organ
Electron	Chromosome	Organ System
Neutrino	Cell nucleus	Organism
Wave	Cell membrane	Human artifacts
Atom	Mitochondria	Opinions and ideas
Antimatter	Cytoplasm	Individual life forms
Ion	Other cell components	Family
Isomer	Gametes	Peers
Isotope	Life forms	Organizations
Inorganic compound	Virus	Corporation

Forms, continued

Military	Mesosphere	Sun
Universities	Thermosphere	Solar System
Community	Ionosphere	Proxima Centauri
Town / city	Exosphere	Star Cluster
Nation	Magnetosphere	Dark matter
World Community	Moon	Orion arm of the Milky Way
Human culture	Venus	Milky Way
Civilization	Mercury	Andromeda Galaxy
Earth solid core	Asteroid	Local Group of Galaxies
Earth outer liquid metal core	Meteorite	Galactic clusters
Earth mantle	Comet	Galactic superclusters
Tectonic plates	Mars	Great walls of galaxies
Earth crust	Neptune	Visible Universe
Hydrosphere	Uranus	P-Brane (Theoretical)
Troposphere	Saturn	The multiverse (Theoretical)
Stratosphere	Jupiter	

The major processes

Here are the major processes and forces that comprise existence. They are presented only as a listing to provide one with an appreciation of the depth of the processes. Some of these will be mentioned in later essays or in the glossary.

Big Bang	Planetary core formation	Conscience
Cosmological Inflation	Plate tectonics	The Unconscious
Strong force	Hydrosphere activities	Language
Weak force	Atmospheric activities	Education
Electromagnetism	Volcanic eruptions	Sexual intercourse
Gravity	Polymerization	Social intercourse
Quantum fluctuations	Evolution	Thought
Higgs field	Replication	Action / Volition
Star formation	Natural selection	Civilizing activities
Nucleosynthesis	Adaptation	Opinion formation
Dark energy	Genetic drift	Policy making & legislation
Chemical bonding of electrons	Recombination	Economic transactions
Spin	Photosynthesis	The theory of everything
Revolution	Cellular interactions	String theory
Precession	Synaptic connections	Superstring theory
Planetary formation	Consciousness	M-Theory

Some ideas about existence

Cosmology	Catholicism	Synergy
Quantum mechanics	Protestantism	Intelligent design
Classical mechanics	Islam	Reincarnation
Celestial mechanics	Buddhism	Afterlife
Relativistic mechanics	Hinduism	Morality
Copenhagen interpretation	Judaism	Freedom
Forms	Taoism	Destiny
Processes	Sikhism	Justice
Philosophy	Shinto	Capitalism
Astronomy	Jainism	Socialism
Evolution	(All other religions)	Communism
Natural selection	Mysticism	Totalitarianism
Conscience	Existentialism	Aesthetics
Unconscious	Gestalt theory	Sociology
God	Epistemology (Theory of knowing)	Biology
Pantheism	Ontology (Theory of existence)	Psychology
Vitalism	Chance	Chemistry
Chaos theory	Determinism	Neuroscience
Infinity	Design	
Eternity	Free will	

What are the recurring forms?

Recurring forms exist for a relatively long period of time and greatly influence and affect other forms and processes.

Recurring physical forms: clusters of galaxies, galaxies, stars, planets, solar systems, magnetosphere, atmosphere, hydrosphere, geosphere, electrons, protons, neutrons, chemical elements, chemical bonds, molecules and certain particles.

Recurring biological forms: DNA, RNA, the zygote (germ cell), differentiated cells in the body, tissue, organs (such as brain, heart), neurons, dendrites, axons and the entire biosphere.

Recurring social forms: civilizations, nations, institutions, law, groups and associations, individuals, families, parents and peers.

Recurring personal/psychological forms: habits, emotions, character and personality types; as well as other traits that recur regularly throughout the history of humankind.

RECURRING PHYSICAL FORMS

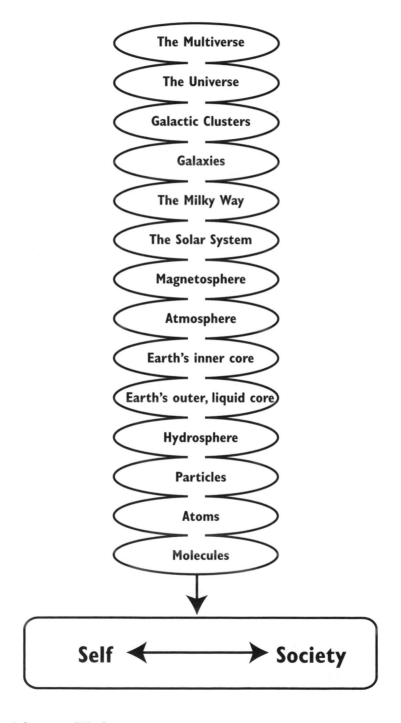

The Multiverse

The Universe

Galactic Clusters

Galaxies

The Milky Way

The Solar System

Magnetosphere

Atmosphere

Earth's inner core

Earth's outer, liquid core

Hydrosphere

Particles

Atoms

Molecules

Self ⟷ Society

RECURRING BIOLOGICAL FORMS

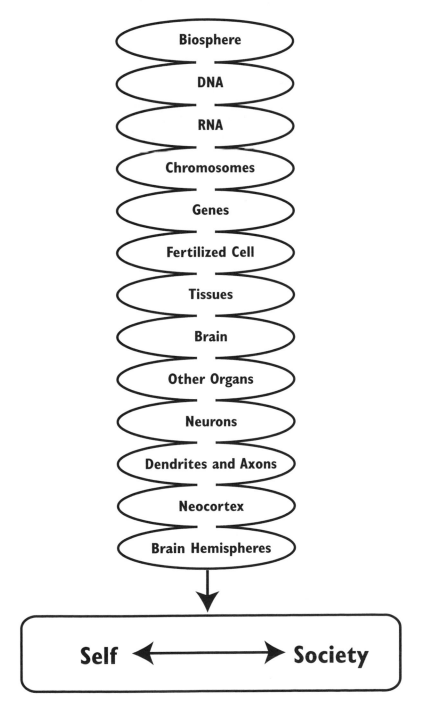

RECURRING PSYCHOLOGICAL FORMS

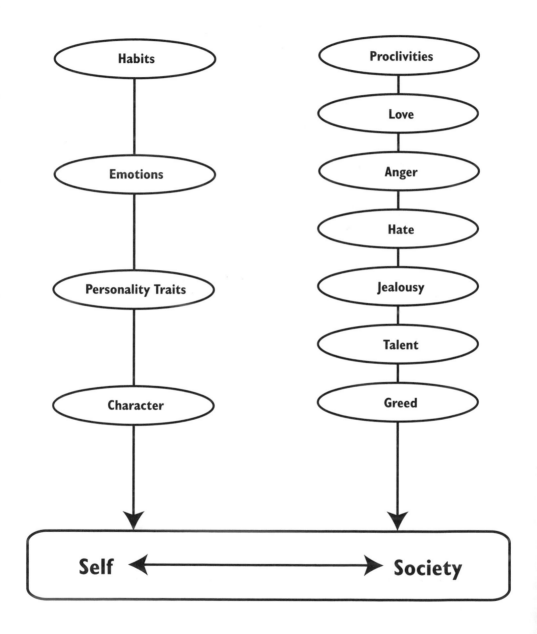

RECURRING SOCIAL FORMS

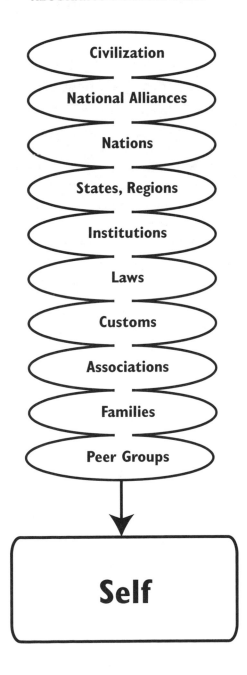

What are the recurring processes?

Recurring processes exist for a relatively long period of time and greatly influence and affect other processes and forms.

Recurring physical processes: quantum mechanics, classical mechanics and relativistic mechanics that include gravity, electromagnetism, strong force and weak force; nucleosynthesis; chemical bondings; spins and charges of particles and atoms; celestial mechanics (including rotation, revolution and precession of moons and planets); and more.

Recurring biological processes: evolution, natural selection, cell division (meiosis/mitosis), cell differentiation, energy production and cell transport, breathing, digestion, central nervous system processes and more.

Recurring personal & psychological processes: consciousness, conscience, the unconscious, cognition, memory, perception, motivation, language, creativity, decision making, volition, judgments, day-to-day chores and more.

Recurring social processes: socialization, education, law making, physical and social intercourse, play, conversation and more.

RECURRING PHYSICAL PROCESSES

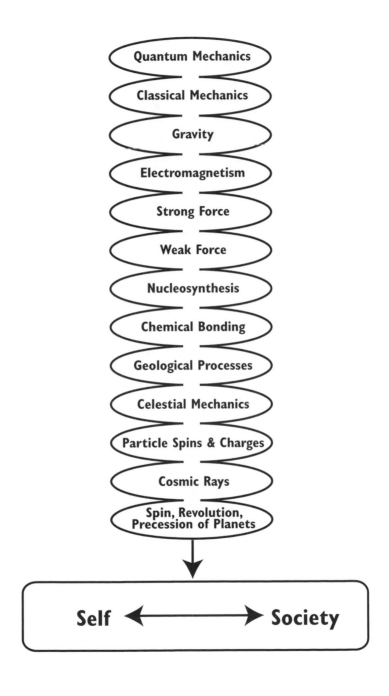

RECURRING BIOLOGICAL PROCESSES

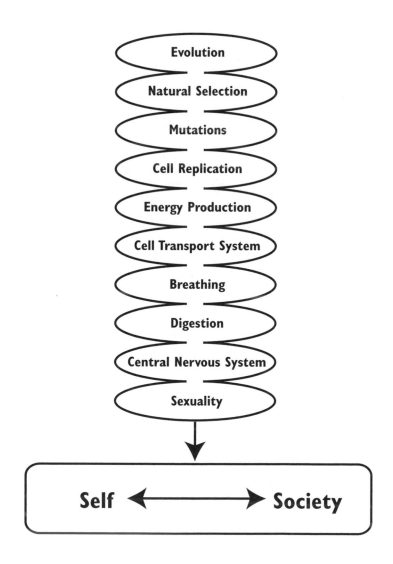

RECURRING PERSONAL & PSYCHOLOGICAL PROCESSES

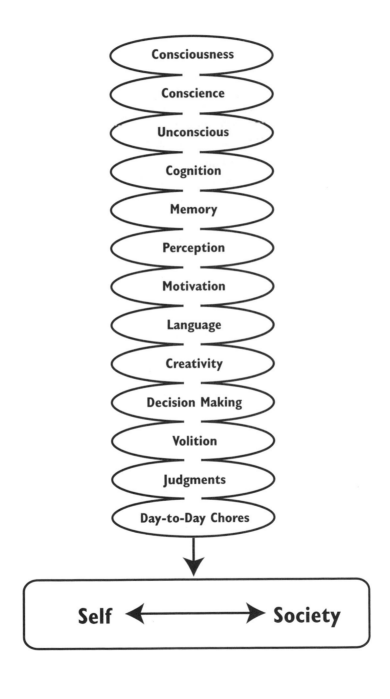

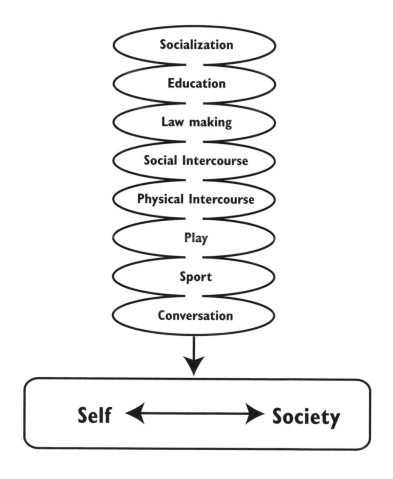

The search for adequate wisdom

Shaped by the incredible array of processes, structures and ideas, we humans have been given the remarkable ability to think, to absorb knowledge and propositions and to make assessments about ourselves. We have a mind that permits us partially to understand the world and attempt to make wise decisions while part of it.

We also have a body and a brain over which we have a limited amount of control. The body and brain have their own reign over us, and while pharmaceuticals and other substances can alter and/or improve our health, perception and mood, the body and brain beat to their own drummer.

But the mind transcends the brain and body and permits a limited free will and thus the ability to seek an adequate wisdom about the nature of existence and life. The mind represents a remarkable synergy of forms and processes, such as the melding together of hundreds of neuron complexes in the brain, along with a vast array of neurotransmitters (chemical and electrical impulses) throughout the central nervous system that trigger an unimaginable array of events that allow us to think and act.

This marvelous mind has led to countless feats, both physical and mental, which have produced all the artifacts abounding on Earth and in near-Earth orbit. The mind has created art forms, literature, music, philosophy, mathematics, astronomy, chemistry, biology and all other branches of knowledge and human achievements.

The mind is the admixture of wholes, including the body and its organ systems, the genes and the neurons; as well the vast array of external forms and processes affecting each one of us, all of which are mediated by the mind.

The more comprehensive our view of the world, the greater becomes our understanding of it. We owe a great debt to the historical events that have lead to our birth and our functioning as thinking beings. And we must acknowledge these influences, past and present and into the future, before we establish an adequate wisdom. Some of the future can be shaped by policies we make in the present, and if these policies are based upon wise choices, the future can be brighter and more productive for humanity and the planet.

The search for wisdom begins with the broad overview of physical, biological, personal and social forms, processes and ideas. We examine how these apparently disparate phenomena actually relate to one another, so that the omission of any of the significant forms and processes will ultimately lead to half-truths.

An adequate wisdom recognizes and highly values the independence and the coalescence of all forms and processes. This means that each form and process has its own uniqueness and yet participates in the assemblage of other forms and processes to form a synergistic whole.

Astonishing observations

Sometimes we neglect to concentrate on the obvious, taking life for granted and moving forward in the hustle and bustle of daily routines with only a glancing nod to rather startling observations, chief of which is that we exist! What could be more astonishing, more astounding than our very presence in the world?

The more we dwell on the wonder of our existence, the more we begin to ask questions about life and the universe, the closer we come to developing integrated ideas about the world.

Not only do we exist, but so do some nine million other species that share this small spinning body we call Earth. Remarkable life forms fill every nook and cranny on the planet. Each species is built upon the astonishing cell that carries instructions for the formation and function of all living things. The microscopic cell is such a complex factory of organic molecules performing so many varied and complex tasks that anyone studying the cell (especially the neuron) must be in a constant state of awe.

If that were not astonishing enough, our home planet is but a speck in the unimaginable universe, whose size and true nature defy human comprehension. The solar system in which we find ourselves occupies an infinitesimal position on the edge of one spiral arm of the Milky Way Galaxy, our home galaxy which contains perhaps 200 to 400 billion stars, one of more than 150 billion galaxies in the observable universe. Our galaxy is so enormous that it takes light about 100,000 years to cross its length! (Each light year equals about 5.88 trillion miles or some 10 trillion kilometers.)

The galaxy is so large that it takes 250 million years for the Sun to revolve around it, moving at about 140 miles per second (or 225 kilometers). The largest galactic neighbor to the Milky Way is Andromeda, some 2.5 million light years away and over twice the size of our galaxy. It is estimated that the Andromeda Galaxy may contain nearly one trillion stars.

Most astonishing among all the amazing features of the cosmos is the fact that from a very, very, very tiny point, a big bang and inflation occurred to create a stupendous universe, the size of which we cannot even begin to grasp.

And not meaning to throw a wet blanket over all these astounding observations, what could be more agonizing than the realization that our own time spent in this remarkable world is limited and that for each and every one of us someday our life will end, as it has for all living things preceding us. Whether or not some transcendent form exists after death is the subject of much debate and speculation.

Restating the tools in the search for adequate wisdom

Let us restate the cognitive tools that provide guidelines in our study.

Forms, processes and ideas attempt to summarize all of perccived existence.

Forms are all the matter and structure in the universe, from galaxies and stars to chemical elements, living cells, life form and social institutions.

Processes are the interchanges between and among forms, from cosmological, biological and social evolution to the workings of the neurons in the brain to the rotation and revolution of the Earth and all other celestial bodies.

Ideas are mental constructs which allow us to think about forms and processes and make observations about them, as well as establish policies based upon such ideas.

We cannot actually separate forms, processes and ideas, as they operate interdependently.

The modes of existence influence all the forms, processes and ideas. They are: design, determinism, chance, free will and synergy.

By design is meant the construction of plans and policies to create new or modified forms and processes.

By determinism is meant that which causes other things and processes to exist and function.

By chance is meant unexpected, unpredictable events arising from the random interplay among the countless structures and processes.

By free will is meant the ability for human beings to think, formulate ideas and, using volition, build past, present and future civilizations. Free will and design often operate in tandem.

By synergy is meant the remarkable tendency of lesser wholes to form greater wholes which take on different characteristics, pointing to the hierarchical and holistic nature of existence.

Stumbling blocks to adequate wisdom

Aside from the realization that we can never reach a full knowledge of existence, there also remains the annoying fact that the mind is greatly influenced and flavored by sensuality, personality, habit and self interest.

This is the egocentric problem that can at times restrict the search for adequate wisdom. Confusion and dissonance arise when decisions are made by people whose character is based solely upon the search for all or some of these goals: power and wealth; the constant need for approval; the obsessive pursuit of sexuality; and the stubborn belief in half-truths and non-truths. Thus we constantly face the swirling and bewildering emotional issues that block wise decision making.

The mind can be the source of destructive and harmful actions, especially under an unfortunate confluence of emotions, personality, ignorance and prejudice, as well as dissonant external circumstances.

The mind can create fantasies about itself. It can become obsessed or compulsive about certain desires and needs. An appeal to logic and rationality is often blocked by the power of emotions and sensations.

In order to approach adequate wisdom, one must attempt to cleanse the mind of preconceptions and emotional static or at least recognize the many negative internal and external influences and then rise above them, which is easier said than done, of course, but quite worth the effort.

Nagging questions

At some point in the lifetime of a rational person come nagging questions. What am I doing here? What is my purpose, if any? What goals should I set for my future? How can I create an inspiring, realistic and successful life plan? How can I begin to understand the world around me? Why do good and bad things happen to me or to those whom I love? How can I accept my own death?

Just as we would not travel to a distant location without a map or knowledge of the location, so too would we not want to create a life plan or simply live a life without trying to figure out the many variables that influence us and make us what we are.

Our initial ideas about the world may be off-center or completely inadequate. We may wish to live a life based upon our "gut" reactions that originate from the instincts and emotions rather than the mind.

That's the major rub: the struggle between the mind and the personality and emotional traits. Often times it is said that our emotional nature is stronger than our rational nature. There is truth to this observation. However, depending upon the free will and volition of the individual, the mind can discipline or control some aspects of the bodily and emotional structures within each of us, as well as the ongoing stream of external influences.

The mind and the minds of those in society can produce policies that lessen the strength and power of emotional static. Some of this can be accomplished by more enlightened education. The self can seek controls over emotional pressures by directing more energy and time into solving some of the nagging questions by using the rational part of the brain. Some studies have shown that the practice of meditation may alter brain chemistry and produce positive feelings, such as compassion, attentive behavior and greater happiness; as well as a stronger immune system.

If we can harness the power of emotion and the thoughtfulness of reason, we are on our way to establishing a useful and productive life plan.

Knowing one's place in the universe

There exists an ongoing interplay and competition between the self and the rest of the world. Each adult member of the population exhibits its own particular personality, character and self-esteem and yet must navigate through the other forms and compete for recognition and freedom of action. A rugged and strong-willed person takes actions that will enhance his or her standing in the world. A few human beings can literally create their own "kingdoms" on Earth, amassing great wealth and reputation. Some may become leaders or even dictators, while others become renowned for their intellect, talent and accomplishments, thus achieving great admiration.

Many members of the population who struggle for success and accomplishment are often held back by a poor self image, or a lack of talent and will power or by other unfortunate circumstances preventing their ascent to greatness.

But in all cases of competition between the self and other forms, each person must understand his or her place in the world. No man, no woman is an island. No island is an island! Whatever our degree of accomplishment and success, each one of us must acknowledge the variety of forms, processes and ideas within and without that help to make us what we are. Once we do that, we can operate under the guidance of adequate wisdom and make wise and compassionate decisions and policies.

The reason we must know our place in the universe is obvious. While we exist as unique forms, we did not appear out of thin air. There were and are and will be an incalculable number of forms, processes and ideas that mold us into what we are and will become. Even though we can stand tall as an individual form, we are born here on Earth thanks to the evolution of our species DNA code, the sex cells (gametes) of our parents and the team of medical personnel who helped to deliver us. We are raised by various family groups, whose circumstances have a great effect upon us. We are born into a particular society that may or may not promote individual freedoms.

We are born at various times and in various places on Earth. The universe, the solar system, the biosphere and the social sphere are different at every moment from birth to death, notwithstanding the ongoing recurrence and influence of certain forms and processes. The evolution of the cosmos helps us to understand how and why we exist at all. The evolution of the stars and galaxies throw light on our presence here. The evolution of life on Earth and the evolution of our species, Homo sapiens, provide an understanding of our place in the universe.

So too must we understand the roles that chance, determinism, free will, design and synergy play in our individual history as well as the history of the universe.

Further, we must recognize and appreciate the role that social organizations play in the lifetime of each individual human. We are born into a series of social structures that help account for our development as mature adults. The individual constantly competes with the social forms and processes abounding in one's lifetime.

To state it as plainly as possible, each person's place in the universe is contingent upon all other forms and processes. While we may believe that we are completely in charge of our lives, we are not. Through fortunate occurrences and the use of free will, we may overcome many roadblocks and obstacles, but we resonate with all synergistic forms and processes of the universe and owe our existence to them.

What is the status of the human condition?

The first part of the 21st century featured rampant terrorism, civil wars and national upheavals, as well as near-catastrophic economic disarray around the world; disappointing those who consider themselves optimistic about the progress of humanity. In addition, violent and extraordinary storms, weather patterns and quakes have occurred around the Earth.

Caste systems, subjugation, ignorance, terrorism, prejudice, authoritarianism and competing and often warring religious and political factions continue to exist in the present-day world.

Today's world continues to include less than desirable individual personality traits and emotions that have recurred throughout all human history, such as criminality, lust, greed, hate, jealousy and deceit. (Along with these disturbing traits, a host of beneficial traits recur, such as kindness, joy, inquisitiveness, humor, love and creativity.)

As for a status report on the human condition, let us first look at the dark side.

THE DARK SIDE OF HUMANITY

Throughout all history, we can point to untold millions of people who have been killed or maimed in wars, conflicts or genocides springing from territorial and political disputes; religious crusades and intolerance; and from the evil deeds of despots, psychopaths and ego maniacs.

We have witnessed the social and economic decay in many inner cities, where crime and danger occur daily. We observe the tragic lives of several billion people living in poverty and disease.

We can also point to the heartless behavior of those who have denigrated or derided people who are different from the mainstream, people whose physical or mental disabilities draw scorn and derision rather than kindness and compassion. Certainly there is a nasty part of the human character spectrum that takes pleasure and gain from people's troubles and sorrows.

Also from the dark side of humanity comes a wide range of prejudices based upon fear and ignorance of those who are different by way of skin color, ethnicity, age, gender, physical appearance, sexual orientation and country of origin.

Most nations and cities harbor enclaves of separate races and economic groups which foster mistrust and social inequality. Only minimal efforts have been made to alleviate the unfortunate conditions of so many members of our species.

The unconscionable disparity between wealth and poverty continues to grow, and quite frankly points to the shameful and reprehensible behavior of our political and economic policy makers.

We also face a human population that is for the most part undereducated and subject to whims and opinions based upon misinformation and emotional static. Universal policy design to educate and inform the population does not exist. Rather, from the dawn of civilization to the present, a haphazard approach to public policy has occurred, more based upon chance and self-interest than any thoughtful, longterm design.

Another problem exists in the day-to-day activities of humans. We are for the most part a fearful species, frightened of strangers and skeptical of the intentions of other people. The over-balanced news coverage of crime, deceit and bizarre actions of a small percentage of humanity has added to our wariness and suspicions. We have weapons, locks and security systems, gated communities, police forces, security guards, chemical sprays and a host of other defensive measures to ward off the dark side of humanity. While understandable, this kind of behavior reflects poorly on our species.

THE BRIGHT SIDE OF HUMANITY

From the bright side of humanity, we have created a rich diversity of art, theater and film, music, literature, performance, science, medicine, social science, philosophy, architecture, engineering, humor, conversation, sport and play—and all the other endeavors that produce a majestic human landscape.

The bright side of humanity has also worked hard to mitigate some of the events arising from the dark side (of natural or human origin), and thus we have individual persons and agencies and governments that try to alleviate the ravages of poverty, poor health, physical disasters and epidemics and all the other miseries to which some people are subjected, despite the fact that these efforts are greatly underfunded.

Although still infants along the path of evolution, we may someday mature into a species that wisely governs our planet, while promoting pleasure and responsibility and strength and compassion.

With nearly 200 nations in the current civilization, where and when one is born on Earth helps form a framework for success or failure in the lifetime of each person. But whenever or wherever a person exists, there is always the chance for an excellent life, one filled with joy, creative thought and compassion for all other living things.

We cannot discuss humanity without understanding how we all are part of an entire universe, and how we are affected by the cosmos and the physical, biological and social world around us.

One final comment is necessary about the continuation of bad and good events. Since the belief expressed in these essays is that a multiplicity of forms, processes and ideas recur throughout history, which are interdependent and influence human life, it is safe to state that bad and good things will always occur.

The various determinants of human personality and emotion are persistent and recur, and so we will not see an end to negative traits, such as greed, lust, prejudice and criminality. We can, however, using free will and human design, create ingenious ways to manage our problems by creating and implementing wise policies and regulations. This not to say that we can regulate away the core prejudices held by people of the Earth, but enlightened and humanistic laws and regulations can provide a series of safeguards for those less fortunate than we.

And, finally, a prejudice-free and universal educational system will help to combat the rampant flow of misinformation throughout all nations. This aspiration will unfortunately have to wait for future humans who hopefully will become better managers of the planet and its populations.

Challenges facing the quest
for adequate wisdom

While we apply the tools to codify the nature of existence, we face some disputable ideas and questions that place roadblocks in our way.

We face mechanistic and reductionist ideas which tend to reduce all the forms and processes of the world into smaller and smaller parts, while espousing various forms of determinism, often times denying or neglecting the existence of chance, free will and synergy. There are those who believe that existence can be reduced to the disposition of particles which govern all our actions.

We face religious ideas, which deduce most truths from a "higher" entity, while inculcating humans with oftentimes rigid ideologies based only upon the writings, opinions and preaching of individual human beings.

We face biological ideas that generally tout chance and determinism as the only basis of life forms in their genetic makeup (genotype). Excluded from their thought process is the synergistic whole of a human being which transcends its genetic directions.

We face some neurological interpretations that claim free will is an illusion, that the creativity of humankind is simply the result of the random complex array of neurons. Thus, they say, for example, the writing of a book that claims free will does not exist is not the product of free will.

We face the physical ideas and problems that address the formation of the universe, the creation of atoms, stars, galaxies and the ultimate fate of the universe. We face a quantum universe filled with uncertainty and probabilities.

We face relatively new theories of strings, membranes and quantum fluctuations that suggest that there exists an infinite number of universes. Quantum theory also implies that the entire universe might be interconnected.

We face questions about right and wrong behavior; the individual and the state; rules to live by; decision making; and the purpose of existence.

Some of these problems can be resolved by promoting a comprehensive view of the world, which calls for the coalescence of all forms, processes and ideas, each of which affects the other in a dynamic framework.

Details and the big picture

The details of form, process and idea and the enormous amount of information they encompass have greatly intimidated many people. The profound difficulties in understanding advanced mathematics, genetics, celestial mechanics, quantum physics, philosophy and many other intellectual pursuits make the study of them quite inaccessible to all but a few experts.

Most humans have difficulty grasping the essence of existence and thus often search for much simpler answers to their core questions. Nothing is simpler than to attribute the entire range of human conditions and the awesome nature of the universe to the existence of gods or a God that creates and/or guides all things. Thus religion has been quite successful throughout the ages in providing explanations, however questionable these explanations might be.

But there are alternative ideas about the world without attribution to a God that suggest how and why things occur. One does not require advanced degrees to begin thinking about the universe and all its incredible parts.

To begin a journey of understanding one must look at an overview of the major components of existence and see how they interact. This can be done, at first, without rigorous attention to the minute details, but rather attention to general concepts. Details can be held in abeyance and examined after knowledge of the big picture has been attained.

A major problem with educating young people is the promotion of details without a prior presentation of the holistic nature of the world. When students observe the big picture, the study of the parts becomes easier for them to grasp.

What is existence?

We view existence as that which is, was, or will exist—the sum total of all that is, was or will be. Existence comprises everything we perceive. This includes the universe, galaxies, galactic clusters, stars, solar systems, particles, atoms, molecules, cells, life forms—and everything else, including civilization, law, families, corporations and individual human beings.

Some members of our population believe that something else may exist outside of existence—beyond our perceptual appreciation. But if we define existence as all that is, was or will be, then we can allow for any theoretical, imagined or incomprehensible existences. But if something exists beyond our perceptual awareness, what hope do we have in ever grasping it?

Existence often is taken to mean the present time, but we cannot rule out yesterday's events and forms. We cannot rule out events having taken place hundreds, thousands, millions or billions of years ago since these have molded us into what we are now. As for the future, while we cannot know what will occur in ten days or in a thousand years, we must also include the future in a definition of existence and discuss possible future trends. By using adequate wisdom, we can establish policies that shape the future and allow for the flourishment of humanity and all nature.

The concept of existence as we perceive it comprises the enormous number of objects, ideas and events; yet these phenomena can be somewhat sorted and classified.

As an aside, imagine what it must have been like during the billions of years before the mind began to observe the universe. Energy (radiations) and matter (particles) formed all kinds of forces and structures that eventually led to the existence of humankind and the creation of ideas about the world. It has been merely a nanosecond on the cosmic clock that thinking beings have been able to observe and comment on this fascinating physical and biological world.

What is the human mind?

One of the brain's complexes is the neocortex, a relatively new addition to advanced living things, and the neocortex along with other brain centers turn consciousness in humans into something spectacular—the mind—which permits thought, language, volition, decision making and speculation. The mind allows for limited freedom so that each person can exert some controls over his or her body as well as perform actions and events in the outside world. Essentially, the mind is our foundation of reason, understanding and introspection.

The mind is the source for originality of ideas and creativity. It is each person's contact with the outside world. A properly functioning and educated mind can lead to intellectual and creative heights where an excellent life can be achieved.

However, it must be emphasized that even the mind has limits—the brain has been structured by DNA; the mind cannot repair all body dysfunctions; and it is affected and influenced by its contextual relationships with other forms and processes external to the individual self. The mind is subject to the powerful influences of the body and its organ systems. The mind is subject to the society in which it finds itself.

As for the unconscious mind, it is most pronounced during our dream states, when consciousness seems to be on hiatus. It also appears to be a hidden repository for a whole range of feelings, such as fears, habits and memories—some of which have been relegated to the unconscious to protect the individual or otherwise remain aloof for reasons we really do not yet understand. The unconscious also appears to have little regard for the spacetime continuum, as it produces images and scenes without regard to time or space. Dream events can be compared to non-linear films, where time and space are jumbled and events at times appear nonsensical.

We really do not understand the relationship between the conscious and unconscious mind; however, we know that our conscious mind is capable of thought and volition, some of which rise above unconscious influence.

How do we know what we know?

Can we trust our senses and brain to provide an accurate assessment of existence? Or does the central nervous system merely provide an approximation of the real world through filters in the brain?

This quandary has always been a robust philosophical topic, and rightly so. There could exist an assortment of unknowns that cannot be grasped by the limitations of our mind. The "real" world may be eternally beyond our grasp.

Because of the central nervous system, the individual becomes aware of his or her own feelings, emotions, thoughts, pain, pleasure and the world at large. Perception is relative in the animal kingdom, with some organisms able to see different forms of radiation or hear or smell with lesser or greater acumen. Humans are privy to a very small slice of the electromagnetic spectrum that produces waves we call light and sound.

Science and technology have created a vast array of instruments that allow much wider penetration into the world, whether it comprises the incredible smallness of particles, atoms, molecules and cells; or the incredible vastness of space and the galaxies and clusters that exist millions and even billions of light-years away from us.

As a practical matter, we trust what we see, smell, touch, hear and taste; and then make assumptions in our mind about the world in which we are both a single individual and part of a larger group of individuals. The common sense approach works fairly well, except when we examine the world of sub-atomic particles or the concepts of time and space.

But for everyday transactions among humans and nature, common sense has proven to be a somewhat reliable and adequate gauge for what we know. While recognizing that counterintuitive phenomena also exist, using reason will help to form an adequate wisdom by which we can successfully lead our lives.

PART TWO:

QUESTIONS & IDEAS

Part Two offers a wide range of questions and ideas about existence, such as determinism and free will; synergy; a goal directed universe; vitalism; chance; mysticism; string theory; and cosmic fine tuning. A key point is raised about the individualism of forms and their inclination to form larger, synergistic wholes. There are listings of the major questions in the world, as well as a discussion of evolution versus preformation or foreordination.

Why are things the way they are?

We are born into a world of spectacular structures and events. Thanks to the interplay of chance, determinism and synergy (and perhaps design), the universe evolved into an extraordinary amalgamation of very small and very large entities, from quarks, protons, neutrons and electrons to stars, galaxies, solar systems and planets. And then, seemingly miraculously, along comes the arrival of life forms, some of whom evolved into intelligent human beings.

We owe our existence to the stars that exploded and emitted atoms that led to the creation of planets and organic life. Many of these atoms formed simple to very complex molecules because of electron bondings.

In living things, atoms combined in complex ways to create molecules which formed long-chain polymers and an incredible variety of forms, such as nucleic acids (RNA, DNA) and countless numbers of proteins and cells, including the neuron complexes in the brain that allow us to feel, read, think and create—thanks to the processes of evolution and natural selection

We exist because other forms and processes evolved to create us. Things are the way they are because of the spectacular interdependence of determinism, chance, design, free will and synergy—all of which affect our biological and psychological makeup and our day-to-day routines.

But there are a great many members of our population who believe we are the way we are not simply because of physical forms and events, but because of an unknowable entity, such as God, which has designed the physical universe as well as the biological domains. This belief allows humans to assign meaning to the physical forces that assembled us and to have faith in something beyond reality.

Dismissing for a moment the numerous dogmas that percolate throughout all religions, we cannot arbitrarily and completely deny the possibility that somehow the universe was "designed" by something beyond our comprehension.

Ideas about preformation cannot be disproved. Thus it is not possible to sustain a point of view that only considers the strict, causal flow of physical evolution, even though this appears to be the most commonsensical explanation available to us. We should all acknowledge that many counterintuitive and possibly unknown phenomena may exist, and thus we must not categorically insist that there is only one paramount explanation of why things are what they are.

What appears to be may not always accurately reflect the way things actually are, and thus we should maintain a healthy amount of skepticism when dealing with the entire universe and the overarching nature of existence.

Determinism and free will

Usually linked to a causal chain of events, determinism states that one thing causes another thing to occur. Virtually all chemical and biological reactions cause other chemical and biological reactions to occur. The mathematical laws of classical physics and celestial mechanics are highly deterministic. So too are the DNA instructions in each cell, giving rise to several thousand molecules that help assemble a wide variety of globular proteins to perform a countless number of functions and interactions.

But other things in the world need not be deterministic. In the bizarre world of quantum mechanics, there is a probability of events rather than a precise determinism of them. Things are not always what they appear to be.

The cognitive and creative processes of a human being, even though influenced by some deterministic principles, can transcend determinism.

Some members of our population believe that only physical laws and mathematical truths are salient. To them, humans are mere physical and biological machines without free will or chance entering the picture. These hard line mechanists fail to see the whole picture. They never discuss the synergistic combinations of forms and processes. To them, only process creates form according to strict mathematical laws.

Some scientists claim that neurological research proves that we have no free will, and yet these same scientists write eloquent books about determinism. What possible assemblage of deterministic laws would initiate the writing of such a book if it were not for free will? It is the mind that creates and manipulates the alphabet to write. It is the mind that uses free will to recognize and manipulate the mathematical laws governing the creation of music or science. The mind utilizes free will to create an amazing range of artifacts—from paintings and books to computer software and high definition televisions.

Because of this capacity to manipulate nature, free will exists.

What is synergy?

Stated as simply as possible, when two distinctive forms combine with one another, a synergy has occurred, producing a new form.

From a broad view of existence, we see an incredible variety of things which can stand alone, as well as interact with other things. We call the things that stand alone wholes—unique collections of parts (which are themselves lesser wholes).

For example, let us look at a human being. While it certainly interacts with other humans and society, it stands alone as an encapsulated entity, with its skin as its protective membrane. The human being is a whole. But then look at the parts of a human—the genetic material (itself many wholes), specific cell groups and organ systems (also wholes), the brain (another collection of distinctive wholes) as well as other entities that comprise the human being.

All the sub-wholes or parts of the human combine to form a whole that is greater than the sum of its parts. *The organism is greater and more distinctive than the DNA and organ systems that produce and sustain it.*

Whenever parts combine to form a new form or whole, the whole is greater than and different from its parts. This is what is known as synergy.

Take six hydrogen atoms and combine them with six carbon atoms in a specific bonding relationship and we get benzene—a completely different whole from its two parts.

A family is a good example of synergy. Parents act quite differently when each one is alone. When they are together, a synergy occurs which is greater than each individual. Add children and other synergistic wholes occur. Humans, like all other structures, tend to form synergies. Obviously, a different dynamic occurs when an individual associates with friends, family, superiors at work, and so on.

Look at a screenplay, actors, lighting, music, photography, settings—all separate factors—yet when combined into a motion picture the whole is certainly greater than all its parts.

The new whole takes on a new persona, so to speak. When we combine an oxygen atom (with its own characteristics) and two hydrogen atoms (with their own specific properties), the formation of water occurs, something which we initially would never have been able to predict, and of course water has its own characteristics, different from its two components.

Other examples include the sperm cell and the egg cell, two separate wholes, whereupon the two form the fertilized egg or zygote, which is a new whole that directs the assemblage of a new living form.

This is the phenomenon of synergy. Physical and biological forms and processes tend to combine with one another. Which begs the question, is the entire universe a synergistic whole that has its own identity? Obviously, we cannot fathom the nature of the assemblage of stars and nebula in a galaxy, nor we can fathom the assemblage of all the clusters and superclusters of galaxies which form the universe. But each greater entity may have a "characteristic" of its own. Which could mean that the universe itself has some special identity (for want of a better word) which transcends all its parts, and which will always be beyond our comprehension.

If an infinite number of universes exists, each universe could be part of a greater whole, and so on ad infinitum. We know that each combination of wholes creates a unique whole, which in turn can become part of a greater whole.

Freedom and free will

To be free of constraints and to be able to think and act according to one's own beliefs –this constitutes the idea of freedom. The degrees of freedom vary from civilization to nation to political regime to social constraints (law and custom) to institutions, to families, to peer groups and associates.

There are those who believe that freedom means we must avoid all or most governmental regulations; but in a synergistic, holistic society, people cannot simply fend for themselves at the expense of others. Fair government regulations are required to operate a responsible civilization which protects the environment from entrepreneurial avarice and which offers opportunities to those who have been adversely affected by circumstances.

No one can ever be truly free. We know that determinism and chance play a large role in the development of a human being, and thus being completely free is a goal, not an actuality.

Free will is generally regarded as the ability to direct one's own lifetime while negotiating the demands of powerful entities, such as peer groups, families, governments, corporations, religions and other institutions. But it must be pointed out that free will may also include irrational decisions from a mind that is buffeted by misinformation and prejudice.

There are very strong influences upon a person that direct or guide one's actions. Circumstances of the parents, the genetic make-up of the person, the central nervous system, the time and place of birth, the laws of the land, the peer groups and countless other influences restrict free will.

But free will flourishes in the mind, where ideas transcend many of the otherwise restrictive forces influencing humankind. Thinking and decision making are examples of such free will.

The goal-directed universe

Design—the planned creation of structures and forms—suggests that there may be a goal which directs all events occurring in the world. This has been called the teleological argument. In mundane affairs, the goal could literally be reaching the end zone in American football, for which an entire set of rules or designs have been developed to reach that goal.

We can say that the acorn is teleological because it contains all the necessary design features to create a magnificent tree.

In a cosmic sense, design could come from a Maker or God, or it could be built into the very fabric of existence and spacetime. Or it could be none of these. Perhaps no such goal exists, at least in the physical world. Without the design mode, then we are left with chance, determinism and synergy to craft a universe filled with the bounty and diversity of forms and processes.

In the biological, personal and social frames of reference, goals are plentiful.

In the biosphere, the goal of all living thing is to survive and procreate. Survival goals include obtaining food and shelter; while procreation goals include finding a sexual partner (and all the hoopla that goes along with sex in higher level animals).

During biological evolution, the goal of creating new species has been mostly gradual, but occasionally punctuated by "sudden" eruptions of new species (such as the Cambrian explosion which produced the structural forms of a majority of advanced living creatures over 500 million years ago.) This suddenness resulted, in part, from chance events in the environment and the geology of the Earth. We are not certain about other events that brought about this grand array of new life forms.

For a human being, goal-direction becomes a major part of one's lifetime, such as creating a viable and flexible life plan; seeking approval; achieving sexual gratification; appreciating the arts and humanities; graduating from school; seeking and sustaining friendships; finding a job and a mate; raising a family; devising investment strategies; forming opinions; and developing hobbies and talents.

In the social sphere, goals are established continuously by laws, customs, requests, demands, mores, official notices, managerial edicts, sales targets and a whole host of proclamations, creeds and game plans.

Vitalism

This is the belief that life is ultimately not reducible to the laws of mathematics, chemistry, physics and biology, but rather some indefinable force that directs the evolution of life forms. An extended version also claims that all forms and processes emerge from an unknown directive that transcends the physical and biological sciences. We might also add that the stars and galaxies could represent more than mere physical aggregates, that they somehow are "alive" with an essence beyond anyone's comprehension.

This latter interpretation blends some aspects of the concept of animism, whereby all forms and processes possess a world essence, for want of a better phrase.

Of course most people refute any such notion. They believe that what can be measured is real, and that all other ideas lack any credibility. And so it is a very hard sell to convince the mind that physical objects and processes have a "life" of their own even though we acknowledge their influence upon us.

But some further examination is called for here. If we were to believe that somehow a preformation principle exists, then design becomes a possible answer to why forms and processes could be merely steps to a pre-concluded end. This may support the idea that the "chicken" may precede the "egg" –as counterintuitive as it may sound.

This is not to say that large physical systems have any kind of "essence" that would or could be understandable to the human mind. But we can at least offer the possibility that galaxies and galactic clusters may not be merely random assortments of nuclear reactions, black holes and electromagnetic spurts; but rather could somehow represent something with an essence beyond our comprehension.

In a thought experiment, imagine yourself as a single tiny observer situated inside the human brain with its 100 billion neurons, and you have the ability to think. You have an instrument that allows you to observe the firings of innumerable cells and the countless interactions among brain complexes. You take pictures of the hundreds of neuron complexes and try to study them. But you would not be able to discern why this organ is doing what it does, as you are merely one very tiny observer incapable of seeing the big picture, not knowing what this object is and not knowing that it is part of a larger whole. *To you, all the brain's astounding activities would suggest an elegantly dynamic but random assortment of physical events with no discernable meaning, just as a galaxy or the universe appears to us.*

We draw a very firm distinction between physical systems and biological forms, but could the entire universe act as a super organism? Quantum mechanics seems to suggest that the universe may be bound together by many different fields which permeate the spacetime continuum. A "living" universe is quite a stretch of the imagination, but remains a possibility nonetheless.

Much of what we have stated is speculation; nonetheless, some of the principles of vitalism are worthy of examination.

A chance only existence?

There are those who believe that things spring from other things without any design or purpose at all. For example, the explosions of stars give rise to the elements that create living creatures on a planet situated at the perfect distance from a star, a planet which contains water and other features that by chance give rise to intelligent living things. The concept also holds that mental events in the brain are the result of physical events and therefore rule out any free will. Thus existence may be merely an epiphenomenon and therefore without any meaning.

It appears as if the Earth is among a hodgepodge arrangement of spinning bodies brought about by chance and determinism and just by sheer chance gave birth to life. We are searching for Earth-like planets. But can intelligent life be nothing more than a very rare chance collection of events?

There are an incredible number of specific events that have occurred for intelligent life to have been formed on Earth. How often could these same events be repeated throughout the Galaxy?

It is certainly recognized that simple organic molecules exist throughout the universe, but to have them evolve into human-like creatures might be rare indeed. However, we are searching for planets in the so-called "Goldilocks zone" where eventually we might discover a rocky planet situated at the right distance from a star which may indeed have water and living matter. Thus by chance alone, perhaps new life evolves elsewhere.

Given that there are perhaps 200 to 400 billion stars in the Galaxy and perhaps more than 150 billion galaxies, one would think that there may be intelligent life forms somewhat like us throughout the universe. But many scientists think the number might be smaller than we imagine. Of course, in an infinite universe, all kinds of living things would be found in any number of universes.

Is this a viable idea? If chance alone brings about other forms, there is little to say of design or free will. We could be merely an afterthought—although an exceptional one at that!

Of mysticism

For the entire range of human thought, there have been many attempts to explain existence as something beyond physicality and rationality using appeals to intuition and ideas that transcend everyday experience. The core belief is that the reality we experience obscures the real essence of being, and the only way to reach wisdom is through many different modes of symbolism, self-denial, intense introspection and forms of self-hypnosis. There exist many ideas about that which is "hidden," such as the soul.

Mystics try to commune with God or an eternal, absolute being or entity. In a broader sense, many people view the universe as mysterious, and one can sympathize with this belief. On the large scale, we really are not certain how energy and matter came into existence; we really do not understand the incredible nature of galaxies and dark matter and dark energy; we do not know if the universe is finite or infinite.

The adequate wisdom we seek must admit to the shortcomings of our knowledge, and while we can sort through the various forms and processes to make educated guesses about how the world works, there is always the nagging suspicion that we may never have it quite right. Thus we must remain skeptical of all ideas, including the mystical approach to existence as by its very nature is practiced without regard for the rational mind.

Many members of the population claim to be of a spiritual nature, which can signify a variety of ideas or feelings, including the belief in a higher power or simply the belief in the goodness and fairness of humanity. Following a spiritual life does not necessarily negate the recognition of forms and processes.

Adequate wisdom and the
modes of existence

Let us first look at the modes of existence as separate entities and discover what wisdom can emerge without an interdependent or holistic viewpoint. Then we will consider the mode of synergy and its integration of forms and processes.

ADEQUATE WISDOM BASED SOLELY ON DETERMINISM

If determinism alone is responsible for all that exists, if the universe and all forms and processes within it, including human actions and ideas, are the result of an assortment of atoms and molecules assembling together by strict mathematical, physical and biological laws—then any system of morality or policy-making lacks credence. Free will cannot exist within this purely deterministic, mechanistic universe. There can be no adequate wisdom. All is determined and every human action is therefore amoral. There is no meaning to life in this mode, except for that which has already been determined.

ADEQUATE WISDOM BASED SOLELY ON CHANCE & RADOMNESS

If chance events alone are responsible for all that exists, then the universe and all its parts have assembled into an admittedly amazing spectacle of physical, biological and social forms and processes without any rhyme or reason. In a solely chance-based universe, human thought and actions are simply the result of a series of electrical and chemical interactions in the brain. What we call free will is nothing but a manifestation of brain activity. There can be no adequate wisdom. Again, all actions are amoral. There is no meaning or purpose to existence, except that which has emerged through chance events only.

ADEQUATE WISDOM BASED SOLELY ON DESIGN

If all of existence were designed by an unknowable entity or set of mathematical and physical laws, then determinism would take over; there is no chance, no free will. Even synergistic combinations of forms and processes would be pre-programmed. Cosmic and biological evolution would have a design model, by which the human brain would incorrectly assume it has the ability to make decisions, to make policy and to think on its own.

ADEQUATE WISDOM BASED SOLELY ON FREE WILL

In this egocentric viewpoint, the rugged individual is supreme and in full control of his or her lifetime. The self is master of all he or she surveys. This pompous individuality can also be true for human groups and institutions, so that real free will is subjugated by an ironclad will, of the person, institution or state for egoistic purposes. In practical terms, wisdom based solely upon free will negates the other influences in the world and produces a very narrow mind which, nonetheless, believes it is superior. A tortured wisdom develops here where morality is based solely upon misguided and self-centered viewpoints, which lead to arbitrary rules and regulations. Life has meaning only for the egoist who believes he or she is master of all events.

ADEQUATE WISDOM BASED ON SYNERGY & INTEGRATION

When we examine the structures and processes that comprise existence, we see that there are both individual parts and collective wholes. Individual human beings form families, peer groups, communities, institutions, nations and the world community.

For each human form, the individual has significance and worth, with consciousness and free will resulting from the synergistic assembly of body and brain. Each individual has an inner life, as well as an outer one that is presented to the world. Each individual human being represents the collectivity of DNA, organ systems and the central nervous system and these, together with all the environmental factors, form a unique synergistic whole.

Is the individual form less worthy, less important than the collection of forms of which it is a member? Of course not, since without the individual there could be no greater and different whole. What is meant by greater is simply larger, not better. When larger wholes (like institutions or nations) begin to treat the individual as a less worthy entity, then we enter the road to tyranny.

Adequate wisdom collects all the seemingly disparate parts of existence and attempts to establish a clearer understanding of how the world works. As such, forms and processes must be examined not only separately but as a synergistic combination. Design, determinism, chance and free will must be integrated into a meaningful whole, with all the dynamics operating interdependently.

This is the path to adequate wisdom. The synergistic view of existence leads to the establishment of moral absolutes based upon humanistic principles. Policy making and judgments should reflect the interconnectedness of existence. All variables of existence must always remain in play. Moral judgments can be made and polices constructed to promote the advancement of humanity. Lifetimes have worth, significance. Meaning arises from the holistic expanse of existence.

What are the major questions in the world?

Where does one begin to seek out the questions whose answers inform our thinking processes? In this listing, we will start at the beginning of the evolution of the universe and establish a series of questions proceeding in a linear fashion. This path finds that our questions move from the physical, to the biological, to the personal (or human) and finally to the social structures and civilization.

QUESTIONS ABOUT THE PHYSICAL NATURE OF EXISTENCE

What is existence?
How did existence begin?
Is the universe unique?
Are there an infinite number of universes?
Is eternity possible?
What is the Big Bang?
What is cosmic Inflation?
What is spacetime?
What is string theory?
What is M-Theory?
Does God exist?
What is quantum mechanics?
What is classical mechanics?
What is relativistic mechanics?
What is celestial mechanics?
What are particles?
How did particles come into existence?
What is the electron?
What are atoms?
What are molecules?
How are atoms bonded to one another?
What is gravity?
What is electromagnetism?
What is the strong force?
What is the weak force?
What is entropy?
What are spin, revolution and precession?
How does a star form?
What types of stars exist?
How does a galaxy form?

What types of galaxies exist?
What forms and processes exist within a galaxy?
How will the universe evolve?
How does a solar system form?
How did planet Earth form?
What is the evolution of the Earth?
What is Earth's core and outer core?
What is the mantle of the Earth?
What is the Earth's crust?
How did the planets and moons form?
How did the hydrosphere begin?
What is the atmosphere?
What is the magnetosphere?
What are the motions within the solar system?

QUESTIONS ABOUT THE BIOLOGICAL
NATURE OF EXISTENCE

What is organic chemistry?
How did life appear on Earth?
What were the first cells to appear?
What are the parts of a cell?
What are nucleic acids?
What are amino acids?
What are chromosomes?
What are genes?
What are alleles?
What is the influence of heredity?
How are cells differentiated?
How do cells multiply to form a life form?
What are tissues and organs?
What is natural selection?
What is adaptation?
What is evolutionary descent of the species?
What is the history of the evolution of living things?
How and why do species appear in orders and kingdoms?
Why is there such a diversity of life forms?
How does the brain work?
How do the organ systems work?
What is the future of the biosphere?

QUESTIONS ABOUT THE PERSONAL NATURE OF EXISTENCE

Why is a human different from all other species?
What is the genetic makeup of each human?
How is a human conceived and born?
How does the human brain operate?
What is consciousness?
What is the unconscious?
What is conscience?
What is habit?
What is emotion?
What is personality?
What is character?
What is language?
What propensities has a human inherited?
What talents and abilities has a person been given?
What is puberty?
What is creativity?
How does a human engage in thought, reflection and speculation?
What are opinions?
What is sexuality?
What freedom does a human being have?
How does a human relate to the outside world?

QUESTIONS ABOUT THE SOCIAL NATURE OF EXISTENCE

How do humans interact with each other?
What are families?
What are social groups?
What are institutions?
What is the role of the military?
What is the role of education?
What is the role of religion?
What are bureaucracies?
What are states, regional and national governments?
What is the legislative process?
What are laws?
What are customs?
What is the civilization process?
What is education?
What is social intercourse?
What is the nature of physical intercourse?
What is commerce?
What is punishment?
What is altruism?

Synergy and individualism

In an updated version of *Genesis*, we observe the coalescence of forms and processes into synergistic wholes, so
that particles beget atoms which beget molecules which
beget stars and galaxies which beget solar systems which
beget planets which beget geospheres which beget, in some
cases, biospheres which beget life forms which beget human beings who beget ideas and actions which beget social
forms, including institutions, which beget nations, world
communities and civilizations.

While new forms occur from the synergistic combination
of other forms, it is important to note that each of the con-
stituent forms stands alone as an individual entity, creating an apparent duality.

This duality appears throughout the animal and vegetable
kingdoms, where each species exhibits its own characteristics and activities while at the same time is part of
the entire biosphere, often in symbiotic relationships with
each other.

As for present day Homo sapiens there exists a wide variety
of synergies, from families and mates to groups and institutions. So while we are all part of greater wholes, each one
of us stands out as an individual with specific traits, character and world views and opinions.

There are ongoing relationships between the self and other
forms in civilization. The self struggles and/or cooperates
with other wholes and seeks to stake out its own particular
mark in its lifetime. The self knows it must cooperate with
other forms, but some selves fight against control of its
freedom of movement and expression. The self is subservient to greater wholes but still swims against the current at
times to assert itself.

*Forms and processes maintain their own integrity and in-
dividualism while at the same time become part of greater*

wholes. In essence, the form remains unique and yet be-comes a vibrant member of a larger assemblage of forms.

Forms and processes exhibit dual roles.

Genesis updated

Particles	beget	Atoms
Atoms	beget	Molecules
Molecules	beget	Stars
Stars	beget	Galaxies
Galaxies & Stars	beget	Solar systems
Solar systems	beget	Planets
Planets	beget	Geospheres Atmospheres Hydrospheres
Geospheres	beget	Biospheres
Biospheres	beget	Living things
Living things	beget	Institutions
Institutions	beget	Nations
Nations	beget	Civilizations

String theory and the infinite universe

String theory, superstring theory and M-Theory all deal with the theoretical convergence of everything, in which all four forces of nature (including the once elusive force of gravity) are explained by the fundamental one-dimensional strings, which are thought to vibrate in an incalculable number of ways to produce all of the forms and processes. The vibrations of these strings produce mass, electric charge, spin and other attributes of particles.

It is the energy of vibrating strings that produces mass, so that the faster the vibrations the more massive an object becomes. (Mass may also be the result of the hypothetical Higgs particle/field.)

The string theories have advanced to include an M-Theory that allows for a theory of everything in the universe if 10 space dimensions exist along with one time dimension. In addition to one-dimensional vibrating strings, M-Theory suggests that there exist two dimensional objects, called two membranes (usually called two-branes); also three dimensional objects exist which are called three-branes; and much larger dimensional branes are called p-branes.

Strings and branes are able to form an endless amount of combinations, and because of that it is hypothesized that there could exist an infinite number of universes. Large membranes may include our very universe which may collide with another brane in perhaps a trillion years to create yet another universe.

M-Theory attains the mathematical goal of physicists who have always wanted to include gravity as part of the other three forces which have already been unified.

The effect upon adequate wisdom is clearly seen, since in an infinite universe or multiverse, beginning and end and the necessity for a God are not required. Spacetime may have always existed and will continue to exist forever. Under M-Theory, there is no need for a prime mover or God. (Yet, for semantic purposes, we could call everything that exists *God*, for want of another word.)

What are some other ideas about existence?

DETERMINISM, GOD & THE INFINITE UNIVERSE

Those espousing a mechanistic or deterministic nature of existence posit that there is always something that causes something else to exist, to move, spin and function. This causal viewpoint often advocates a strictly linear evolution with a specific beginning.

In religion, causality is often used as a proof that God exists since it is assumed that every event has a prior cause and if you go backward (to an infinite regression) one would assume that there is a first cause or God that sets everything in motion. Is it a God that created the first particles and fundamental forces? Is it rather simply randomness or chance? Or could it be merely a continuation of events that have occurred eternally and will continue to recur infinitely?

Quantum mechanics has proposed the amazing possibility of an infinite number of universes.

In an infinite universe or multiverse, there would not necessarily have to be a first cause or God, since the notions of beginning and end may have no meaning. This concept is difficult to grasp, but the idea of a beginning and ending may have no cosmic relevance. To begin and to end are constructs of our mind, and in the mundane world have much relevance. But in the unfathomable infinite universe, there may not be any prime mover or initial starting point.

People will ask, but what started everything? Surely there must be something that began everything? And if you answer that God started everything, we could ask what came before God? *If the theory of an infinite number of universes is correct, we might posit that spacetime has always existed and will continue to exist forever without beginning and end and without the need for a Maker.*

Because this concept is so powerfully important, it must be restated. On Earth and in our observable universe, things begin and end. Because of that, we automatically assume that existence and the universe began and may someday end. These ideas are of human origin. They may hold no relevance to the enormity of spacetime, which has every possibility of existing without beginning and end. It may be too much for our minds to comprehend, yet things may have always existed and will always continue to exist. In this continuous existence, big bangs, spacetime and forces and forms may be permitted to exist and to change from universe to universe (or within one endless universe).

THE WORLD AS A LIVING ORGANISM

There is a concept that the universe is a living organism. This idea smacks of animism, yet gains some credibility from the quantum nature of the universe, where the theory of quantum entanglement suggests that there may be connectivity throughout the entire universe.

In adequate wisdom it is possible for the universe to exhibit the Grand Synergy, wherein all structures and forces create a new and higher order formation, so that living things, stars, galaxies, clusters, superclusters and cluster walls and voids all represent one incredible synergistic entity, about which we have absolutely no comprehension.

PANTHEISM

This belief denies the existence of a personal God, one that represents or embodies humanity in any way. This idea views the cosmos and nature as one entire entity. One branch views the physical world as the one substance comprising all things and ideas. Another branch views existence as either mental or spiritual. Yet another branch views existence as a duality between the physical and the mental/spiritual substance.

EXISTENTIALISM

Although this concept varies among its proponents, the existentialist view shies away from science, language and objective reality, dealing mostly with the emotional nature of humankind, including freedom, pain, suffering and guilt. Each individual is solely responsible for giving his or her meaning to life and should live that life with passion. Often, existentialists find life absurd, futile and filled with angst or dread. Any meaning to life is brought about by one's consciousness and not by any predetermined external forces. Each person is responsible for actions taken. Generally, despair comes from the knowledge that the only meaning to life is what we ascribe to it.

In the view of adequate wisdom, existentialism suffers from the half-truth syndrome by avoiding the complete picture of life and the universe. It fails to join the rest of humanity in creating pleasure and responsibility, as well as strength and compassion. It literately lacks a whole world view by excluding many of the twelve variables of existence.

COSMIC EVOLUTION

Mathematicians, physicists, astronomers and cosmologists all have a pretty good idea about the formation of our universe, not before the Big Bang, but certainly afterward. From the smallest imaginable point emerged a super-gigantic universe filled with particles, such as photons, electrons and quarks, the latter which formed protons and neutrons and then stars, galaxies and the 92 natural atoms from supernovas. These atoms combined in multitudinous arrangements to form molecules, solar systems and, at least on one planet, a vast array of living things.

For many billions of years, gravity held the large-scale structures together, only later to experience a repulsive force that now creates a rapidly expanding universe thanks to the mysterious dark energy that permeates the entire universe and expands not the galaxies but the spaces between them.

CHAOS THEORY

For those who subscribe to a complete determinism of all forms and processes, the belief is that the initial conditions of any given form and process will dictate the unfolding and evolution of such phenomena in a precise, predictable manner. But what has been discovered is that some systems that overlap each other have an instability, and no matter how precise the measurements, slight imperfections or changes can occur which lead to a different end result than what would be expected. Weather patterns exhibit such chaos, as just very small fluctuations can and do spell out unpredictability. Even the smallest discrepancy between two equal sets of initial conditions can lead to markedly different results.

Thus chance events can and do affect determinism. Had the quantum fluctuations in the beginning of the universe been of a different nature, stars and living things might not exist. Chaotic systems appear in ocean currents, the orbits of planets, the heart beat and blood circulation, the stock market, snowflakes, geographical borders, Jupiter's red spot and much more.

There is even the odd notion of strange attractors in chaos theory which somehow cause systems to veer off track and follow different patterns.

Chaos theory, despite the notion of unpredictability, does claim to show an underlying order to the universe.

MECHANICS

Classical and celestial mechanics are branches of physics that describe the motions of large objects, such as spacecraft, planets and moons, as well as machinery, liquids, solids and gases. Other mechanics cover the microscopic world (quantum mechanics) and the phenomena of particles approaching the speed of light (relativistic mechanics.) Classical and celestial mechanics are very precise sciences that accurately predict the effects of forces upon other objects in the everyday world. Quantum mechanics specializes in the study of particles, atoms and waves. Relativistic mechanics deals with the speed of light and the influence of gravity throughout the universe. It also deals with the conversion of mass into energy and energy into mass.

Quantum mechanics does not deal exclusively with deterministic systems. Quantum mechanics studies a variety of bizarre and counterintuitive relations involving particles, many of which encounter probability and uncertainty.

TAOISM (DAOISM)

Followed by millions of people, mostly in the East, this religion/philosophy speaks of the hidden nature of existence that evolves and influences all things. It espouses humility, moderation and compassion; and it is deeply concerned with humanity and the cosmos. There is an interdependence of all apparent dualities. The male and female aspects of life intersect one another and point to the Way or Tao of all things.

GESTALT PSYCHOLOGY

The wholes of the mind and that of behavior are a focus point for Gestalt psychology which echoes adequate wisdom by looking at the parts and observing the wholes that transcend them. According to this theory, the parts of human behavior are determined by the wholes.

Could the entire universe somehow pre-date the existence of all matter and energy, and then create all forms and processes as a preformational unfolding of time and space? Heady stuff to be sure. *A Gestalt view of the universe would suggest that the whole determines all the parts!*

Could the entire universe be working backward? In physics, particles can move forward or backward in time. Can the entire universe be fooling us altogether and making things appear to exist from past to present to future, when they really work the other way? This concept, of course, is just too much for us to comprehend; but who really can prove otherwise? In any event, it is savory food for thought!

PARTS AND WHOLES

Adequate wisdom points to the interdependence of all forms and processes and thus claims that all things are contingent upon all other things. But in addition to the holistic nature of existence, there is the paradoxical individualism of each form and process. As humans, each of us exists encapsulated as a unique form, contributing unique actions and thoughts to society.

At the same time, we are part of the synergistic whole—Homo sapiens, our species—and therefore are part of a greater whole. This dichotomy appears throughout the universe and remains an incredibly important feature of existence. From quarks to protons and neutrons; from electrons to atoms; from atoms to molecules; from stars to galaxies; from planets to solar systems; from cells to organ systems—we see the pattern of individualism and synergy, or parts and wholes.

Cohesion and diversity

There exist certain forms and processes that maintain the structure of things, without which there would be sheer chaos and randomness. The interactions of particles and atoms and molecules give rise to the vast universe. All protons are the same. All neutrons are the same. All electrons are the same. (Electrons may spin in different directions and change their energy states when moving up and down the orbitals of atoms.) But the relationships among these three cardinal forms vary and therefore create the wondrous diversity of existence. Certain forms, called gluons, hold the atomic nucleus together, providing cohesion and steadfastness.

Thanks to the repetition of the electromagnetic spectrum, we know that specific oscillations yield specific phenomena, so that musical notes are represented by specific vibrations or frequencies of sound waves, or that visible light is caused by specific wavelengths along the spectrum. Different colors exist because of specific wavelengths. Telecommunications exist because precise frequencies in the spectrum carry information.

Without gravity and dark matter, there would be no human existence, as no stars or galaxies would form and no life forming elements would be created.

The DNA molecule is the superstructure that directs and sustains all life forms. And while its mechanism is the same for virtually all living things, it directs different chromosomes and genes for different creatures and produces the rich variety of life forms on Earth from mistakes, errors and environmental factors. A spectacular fecundity of DNA-directed forms existed or exists, such as the fascinating dinosaurs that lived for about 160 million years to the currently existing fungus in Oregon that is over 2300 acres in size and several thousand years old!

In the human world, there exist all kinds of structures that keep people organized or together, from families to peer groups and associations; from educational, religious and military institutions; to the laws and regulations and moral codes prevalent during one's time and space while alive.

But we humans are quite a diverse collection of skin colors, sizes, shapes, unique personality combinations, opinions, beliefs and likes and dislikes.

Diversity is the result of chance and synergistic interactions among forms and processes, so that no two stars or galaxies or planets are the same; that no two species are the same, that no two nations are the same, that no two humans are the same. Diversity also results from human design, so that no two works of art, music or literature are the same.

Chance, human design, synergy and free will provide the diversity in the world.

What is the largest form in the universe?

The largest known object in the observable universe is a cosmological giant wall of superclusters and filaments of galaxies that is called the Sloan Great Wall. Recall that the length of the Milky Way Galaxy is about 100,000 light years. Well, the astounding length of the Sloan Great Wall measures 1.37 billion light years! It is located about a billion light years from our solar system. (Also, remember that each single light year is about 5.88 trillion miles or 10 trillion kilometers.)

It seems impossible to grasp the enormity of this spectacular cosmic structure. There are several of these great walls that we see, separated by enormous amounts of empty space, called voids. This space may not actually be empty, according to quantum theory, which might permit particles rapidly to appear and disappear in this apparent vacuum or void.

What is the smallest size in the universe?

The smallest size that we can imagine is the so-called Planck length, which is astoundingly small, about 1.6 x 10 to the minus 35th meters.

Planck time is judged to be 10 to the minus 43 seconds (the smallest measurement of time).

The paucity of forms

All the forms that we observe—stars, life forms, atoms, molecules, galaxies, clusters—comprise about 5% of the universe. This is an unbelievable statement. There is an entity called dark matter (less than 25% of the universe) thought to help galaxies remain stable; and the ubiquitous and enigmatic dark energy (about 70% of the universe) that seems to create the current expansion of the universe.

There have been recent studies suggesting that there are three times as many stars in the visible universe as previously estimated (now some 300 trillion billion) and thus the amount of dark matter may be less than previously thought.

Trying to grasp the enormous size of the visible universe and the vast distances between galaxies and between stars is only made more difficult when we realize that matter occupies such an insignificant portion of the universe.

Was there a beginning to our universe?

There is consensus about the "birth" of the universe (or our universe), with most cosmologists believing that in a fantastically small fraction of a second and at an extremely small "point" in space the universe experienced a big bang and a subsequent stupendous inflation that eventually accounted for the enormous universe of atoms, molecules, super clusters of galaxies, stars, planets, moons and life forms.

The early universe consisted of energy and radiations converted into particles that were governed by quantum fluctuations—unpredictable gyrations that eventually created huge clumps of matter from which, under the force of gravity and pressure, stars and galaxies appeared, along with scores of key atoms from exploding stars that formed planets and living things. A remarkable number of events had to occur for our universe to form stars and galaxies and subsequently the chemical elements required to make life possible.

We actually know that our specific universe was "born" with a bang and that the remnants of the bang, some 13.7 billion years ago, are still radiating throughout the universe in the form of microwaves with an extremely cold temperature.

The notion of the beginning and the end of existence or the universe is a human concept based upon the fact that we see things—including ourselves—being born and then dying. In the cosmic world, on a universal scale, there may be no real beginning, mainly because our brain and mind may not have the correct perception. In other words, "beginning" is a concept that applies locally but perhaps not on the grand universal scale. Thus, even though the Big Bang occurred, it may have been the result of previous bangs. If this were the case, then we wouldn't have to ask where spacetime came from—because it may have always existed and always will exist. (This may appear as a facile statement, but in an infinite universe, spacetime need not have a first mover or first cause.)

If there is validity to this concept, then to say that the Big Bang was truly a unique, one-time only event may not be correct. If infinity is the correct notion, then our universe did not actually *begin* with the Big Bang but merely *continued* from one phase to another part of eternity, a never ending series of universes; or possibly just one universe always existing and never ending.

In a thought experiment, imagine yourself as a very tiny observer inside the human body. You are so small that you do not see any cells. Suddenly, a cell divides and a new cell is formed right next to you. Might you not believe that something big just occurred out of nowhere that created something new? In this rather loose comparison, the cosmic big bang might be one of an endless number of bangs and inflations occurring over an infinite number of times, just as the replication of cells would appear to you as the tiny observer inside the body.

The concept of cosmic inflation (the immense expansion of space) also provides for the theory that universes are continually born because of the infinite number of quantum disturbances or jitters, each of which gives birth to a universe. (This is admittedly a difficult concept to comprehend. For those interested, a look at books and articles about quantum mechanics, inflation and infinite universes would be advisable.)

So was there a beginning to the universe? The answer appears to be, yes, our particular universe began with the big bang. But it seems quite possible that this was not a one-time event. A unique beginning implies something out of nothing, which defies logic, and so there is the likelihood of a continuous universe with no beginning; that our universe is merely one of an infinite number of universes. And one might surmise that there would be no end, for if the universe ends then *nothing* would exist, which seems to be impossible (at least to the rational mind).

Just to clarify, it may be possible for *our* universe to reach a heat death, essentially ending *our* universe; but there is every possibility that an infinite number of other universes exist now or will continue to be formed forever. An end to everything is simply incomprehensible.

It appears as though existence has always existed!

What is cosmic fine tuning?

For energy and matter in our universe to have formed, for stars to have emerged and for life to have existed, there are many initial conditions that must have taken place in very precise specifications. If the strength of the Big Bang/Inflation were slightly off, if gravity had just a small difference in strength, if the strong nuclear force that binds protons and neutrons together were slightly different, then stars and life would not have evolved in our universe. There are other instances where just a slight difference in force would have prevented life from occurring. The chance that life exists is highly remote when viewed by the spectacular long shots needed to have occurred. In other words, it is very difficult to assume that chance alone created the precise conditions for stars and life forms.

All the fine tunings of the universe suggests that design is a possible explanation of how everything exists. By design, we mean some kind of blueprint or built-in directions, which can be called determinism brought about by very specific physical laws—a design inherent in the very laws themselves. The only substantial counter-argument that offers a non-design explanation is that there could be an infinite number of universes, so that it is quite reasonable to conclude that at least one universe exists with the proper fine tunings for life to emerge, and we are fortunately living in it.

That there are an infinite number of universes is theoretically possible when considering that there may be 10 space dimensions in addition to one time dimension. Seven of the 10 space dimensions are not visible but fold or compact in such ways to create an infinite number of universes.

Recent ideas about inflation suggest that because of a nearly infinite number of quantum fluctuations occurring after the Big Bang, an ongoing number of universes might have been created or still might be in the creation mode.

What it all boils down to is that one either believes that somehow our universe is the only one and it has been designed by some unnamable entity or by a highly improbable set of chance circumstances; or that there is an infinite number of universes, which means most likely that chance and determinism in an infinite universe created the one in which we find ourselves living thanks to the proper fine tuning of physical laws.

The chicken and egg dilemma

What came first—the chicken or the egg? This supposedly simple question poses many questions not easy to answer. It can lead to a circular argument, whereby the egg eventually produces a chicken, but where did the egg come from if not the chicken?

For one viewpoint, we must go backward in time to the first instance of a replicating cell. Several theories have proposed that during the course of a billion years, life began from the synergy of coalescing organic molecules into a strand of RNA that was able to fold in upon itself and direct cell replication. The RNA contained information that allowed for the eventual formation of DNA and subsequently a range of nucleic and amino acids that carried out the function of "living" and dividing itself. If this occurred, then the "egg" comes first and allows for the formation of the "chicken." It must be stated emphatically that we actually have no definitive knowledge about the creation of the first cells.

But, if we consider prior design or some kind of "acorn" effect, it is possible for the form of the "chicken" to have existed prior to the formation of the "egg." Preformation is certainly counterintuitive, but perhaps all the major structures in the universe have a prior, built-in blueprint that creates the assemblage of key events and wholes we think are a result of cause and effect. We usually say that A comes first and then B follows. But could B come first and then A follow?

What an intriguing thought! If we believe that the design of the chicken came first, then we entertain the notion that every major structure and process might be heading toward an end already prescribed! This is the exact opposite of the way scientists and most people think today.

Of course this notion that the chicken comes first is just an hypothesis. One piece of possible evidence that wholes exist prior to the evolutionary process that seemed to have created them is the phantom pain one experiences having lost one or more limbs. Even though the limb no longer exists, the individual feels the limb and claims it still exists and exhibits pain. What this may indicate is that there exist some "fields" (for want of a better word) that act as the blueprint towards which forms gravitate. (Countering this argument is the possibility that the brain maps out all our limbs, and the apparent phantom limb phenomenon is simply the brain's way of making up for the absence of the sensory feedback from the missing limb.)

It's also noteworthy to realize that it is the *shape* of each individual protein in the cell that brings about its function. In a human cell there are thousands of proteins all performing different tasks, and it is the three-dimensional shapes of most of these that dictate the process. *Thus in this instance form initiates function.*

Another possible hint of preprogramming is the embryonic stem cell, which directs the formation of all body parts. No one knows at this point in time what mechanisms are at work to direct the stem cell to create all the specific tissues and organs in the body. Could they be manifestations of some unknown design? There are other changes in the genotype that are not attributable to the DNA code, and these are called epigenetic effects, which soften the ironclad beliefs of most biologists that only chance and determinism are at play in life formation, evolution and maturation.

Photons, electrons and protons exist for an extremely long period of time. The origin of all matter may encompass the chicken and egg question. Did particles and photons emerge by chance or by design? There is a theoretical concept, known as the Higgs boson/field (both a form and a process), which is thought to have given matter its mass. Again, if true, this seems to be some kind of design as well. So, once these particles and forces formed, they may have become determinants of elements, stars, planets, galaxies and life forms. Chance allows for variation of structures; therefore we have a wide diversity of stars, planets, and galaxies, as well as superclusters of galaxies.

As for the incredible number of living species on Earth, it is difficult to believe that design and determinism created them all. It is certainly the case that chance events helped to saturate the Earth with about nine million species, each with its own master DNA code.

We can also ask, does a form exist because of a process; or a process exist because of a form? The interaction of form and process in all wholes seems to indicate that both are viable explanations! Structure appears to dictate process (as in the structure of proteins), while process seems to dictate structure (as in cellular replication.)

Also, it must be pointed out that design and determinism are closely linked. Once something is designed, the steps to follow out the design process are usually deterministic, with chance events always lurking.

The entire concept of preformation is speculative, but we cannot categorically dismiss this unusual theory. If we accept it, we would have to believe that the whole is determining its parts. This makes no sense, unless we are being brilliantly fooled by the universe and thus hold an incorrect concept of time and space.

Entropy

Physical laws state that the universe moves towards an ever increasing heat loss, a concept known as positive entropy. Whether design exists or not prior to the formation of certain wholes or structures, these forms eventually decay, die and transform. Wholes (including humans) are known as temporary negentropic structures (that is, forms that operate under negative entropy.) These self-replenishing forms overcome the effects of heat loss, but as we are unfortunately aware these forms eventually begin to slow down and die. Stars are negentropic while they burn hydrogen into helium and other elements; but they eventually experience a heat death, some quietly, some spectacularly with a supernova.

We cannot dismiss the concepts of past, present and future. There exists a chronology of forms and events moving ahead from a minute spacetime to an ever expanding universe, and along the way, stars and solar systems happened to have appeared, leading to life forms temporarily functioning as independent and collective entities until their eventual end. (Don't forget about the seemingly impossible idea of time moving backward, so that the whole is determining the parts.)

According to M-theory, there is the possibility that a complete heat degradation of this universe does not occur, and that in a trillion years, give or take, matter and energy once again coalesce into a membrane that collides with another brane to form another Big Bang and another universe. If true, this of course would negate the idea that there will be an eventual heat death of the universe, as well as suggest that the direction of time and entropy is not well understood.

If we subscribe to the notion that the world is infinite, then even if certain forms die, other forms will continue to exist until their demise, whereupon other new forms emerge ad infinitum. The tragedy for each living person is of course the reality that each of us will not be part of continual evolution, unless we subscribe to the speculative belief of an everlasting soul.

What is destiny?

Destiny stands for the concept of the predestination of events, that is, things that happen have already been planned, either by God or some other unnamable, unknowable force or forces to reach a specific end result.

Other ideas, like fate, fatalism, or determinism are used to define destiny.

The concept of destiny—the opposite of chance—alleges that there are no unrelated events; that events are governed by determinism or an ineffable "force" guiding physical and biological entities.

While most people believe that free will undermines the concept of destiny, there is no way to know if external processes do indeed create a fate for all things.

Some believe that the synapses in the brain determine our behavior before we do, but since the brain processes are both electrical and chemical (the latter a slower process), we may be able to think original thoughts without complete determinism. The mind is the result of a synergy among body, brain and the external world. Nonetheless, it appears as if we will never know if greater forces are at work completely guiding and defining our actions.

There is also the possibility that the final form of each event or object is determined, but that the way to each final form can contain elements of creativity and free will. This concept of equifinality allows for a predetermined whole or end-state to be reached by many different paths.

Based upon observations of humanity, it does appear as if there may be something to the concept of destiny, that a person's overall fate or character may be somewhat sealed, so to speak, but that the way he or she lives is still somehow affected by chance and free will—akin to the notion of equifinality. Of course this notion is paradoxical, but yet curiously rings true.

ADEQUATE WISDOM

PART THREE:

PHYSICAL EXISTENCE

Part Three provides brief summaries of the major components of the physical structures and forms within the universe, including the particles, atoms, stars, galaxies and the solar system.

Foundations of forms and processes

The foundations of all structures are the electron, the proton and the neutron, all forming a variety of synergistic forms which operate in concert with the physical processes. Such processes include the gravitational force that helps to create stars and allow solar systems to exist; the strong force which holds the atomic nucleus together; the weak force that permits decay and radiation and assists in star formation; and electromagnetism, which is the foundation of the atoms and molecules that constitute all matter.

In addition to the basic physical laws that govern the forms and processes, there exist an incalculable number of other forms and events that give rise to the entire landscape of existence. From black holes and supernovas to planetary magnetospheres; from the three-dimensional folding of organic molecules to the millions of DNA master codes; from synaptic junctions in the brain to social institutions and laws—there exists an astonishing number of forms and processes interacting with one another.

We must also mention quantum mechanics which studies particles like quarks that combine to form neutrons and protons; as well as the relatively new theory of strings, whose hidden compacted sizes, complexities and vibrations might well create all the subatomic forms and universal processes. In addition, a theoretical M-Theory features strings and spectacularly large membranes which may well explain the nature of the physical world and the possibility of an infinite number of universes.

Quantum actions could have initiated a universe eventually allowing for the fine tunings required for life to exist. Perhaps thanks to those specific quantum fluctuations during the inflationary expansion of the universe, we are alive today!

Some theories suggest that these quantum fluctuations or jitters might have spawned an infinite number of universes.

What is a quark?

There are six types of this elementary particle, but we are only concerned with two of these—the up and down quarks, which are distinguished from each other by exhibiting different electric charge and spin, as well as other properties.

The proton emerges from the synergistic combination of two up quarks and one down quark. The neutron is comprised of two down quarks and one up quark.

UP quark + UP quark + DOWN quark = PROTON

DOWN quark + DOWN quark + UP quark = NEUTRON

There are other quarks that are unstable and quickly decay. There are also anti-quarks, as all matter is thought to have its own opposite form.

The proton and neutron become new synergistic forms from the amalgamation of these up and down quarks. The quarks (and electrons) are believed to be elementary and not divisible by any other form, except for the possible influence of superstrings or the Higgs field/particle, both of which are speculative at this point in time.

The quarks are held together by the strong nuclear force, mediated by forms called gluons. Here is yet another example of the complementary interplay between form and process. That is, a force or action (nuclear force) is "mediated" by a form (gluons).

What is a proton?

Based upon the synergistic combination of three specific quarks, the proton is the basis of the atomic nucleus and is a very long-lived particle with a positive electric charge. *The number of protons in an atom determines its chemical property.*

When one proton is combined with one electron (a negatively charged particle), the synergy creates hydrogen.

p+ ----------------- e- = Hydrogen

When a neutron is added to the proton, we have a new synergistic form called an isotope.

p+ n-------------- e- = Hydrogen isotope called Deuterium

When another neutron is added we see yet another isotope forming:

p+ n + n ---------- e- = Hydrogen isotope called Tritium

Just like the quarks, the protons and neutrons are bound together in the nucleus of the atom by the strong nuclear force (which is mediated by particles called gluons.)

What is a neutron?

The neutron is slightly larger than the proton but has no electric charge. It is a synergistic result of three specific quarks. In the nucleus, the neutron is very long-lived; however, as a separate particle it will decay into other particles.

The neutrons and the protons are held together by the strong nuclear force and are most often accompanied by a number of electrons which form fuzzy orbitals about the nucleus. As additional neutrons are added to the same number of protons, various forms of a specific atom, called isotopes, are formed and represent new synergies.

A carbon atom contains six protons, six neutrons and six electrons:

p+ p+ p+ p+ p+ p+ ....n n n n n n......e- e- e- e- e- e-

If one more neutron is added to the carbon atom, we get carbon 13, six protons and now seven neutrons.

If two neutrons are added to the carbon atom, we get carbon 14, a radioactive isotope used in dating archeological and geological structures.

Once again, we see the effect of the synergistic assembly of new forms.

What is an electron?

Other than the possibility that electrons are the result of theoretically vibrating strings, the electron is thought to be an elementary particle, not a composite of any other forms and thus, like the quarks, a building block of all other synergistic structures in the universe.

It is exceedingly small compared to the nucleus of the atom, and its negative charge attracts it to the positively charged proton. The electron makes possible all the major forms in the world. It swirls in a fuzzy path about the atomic nucleus in many different orbitals, creating both waves and particles (photons), as well as generating electromagnetic fields. The electron is responsible for all the different atoms and molecules in the universe thanks to chemical bonding.

The electron provides the energy for all electronic devices, from microwave ovens to computer systems. When coupled with the proton and neutron, it forms one of the most dynamic synergistic forms and processes in the universe.

The electron is the key to consciousness and the central nervous system. When an electron is added to an atom it becomes a positively charged ion. When an electron is removed from an atom it becomes a negatively charged ion. Inside the neuron cell positively charged potassium and sodium ions react to negatively charged ions outside the cell (such as negatively charged chloride ions) and create a difference in electrical potential.

Based upon the structure of the neuron cell membrane, certain ions are permitted to move between the cell and the external chemical fluid, while others are not. The result of all these electrical and chemical interactions permits the neuron cells to function as they do and allows for the propagation of electrical connections over long distances, thus forming the basis for the central nervous system and consciousness.

What is electromagnetism?

One of the four fundamental forces or processes in nature, electromagnetism holds the electrons together in atoms and molecules by electric and magnetic forces. A moving electric field creates magnetism; conversely, a moving magnetic field creates electricity and thus allows for transformers, induction motors, and generators.

Electromagnetism is the foundation of all chemistry and of all life.

THE ELECTROMAGNETIC SPECTRUM

In our pulsating universe, particles and objects radiate oscillations or waves along a specific spectrum as a result of electric and magnetic fields and impart energy when in contact with other matter. The radiations are measured by their wavelengths, so that very long wavelengths impart little energy, such as radio and microwaves. As we continue along the spectrum, we find infrared and visible waves (light); and then farther along the spectrum we find ultraviolet, x-rays, and finally, gamma rays. The latter two are ultra-high frequency waves that can cause serious harm to living things. Ultraviolet (UV) rays from the sun that are taken in moderate doses create a healthy-looking suntan, but extensive exposure to this form of radiation is harmful.

THE UNIQUE CHARACTER OF ATOMS AND MOLECULES

Each atom or molecule gives off its own special emission or signature that enables scientists to identify the form. This proves very useful when examining distant objects, like stars and galaxies, where emission lines show what elements are contained in the form being studied.

What is a neutrino?

Along with quarks and electrons, neutrinos are elementary particles that are pervasive throughout the universe wherever there are stars. Neutrinos are related to electrons though they have no electric charge and very little mass. They are produced during the fusion process of hydrogen into helium which powers all stars and all solar systems. The odd thing about neutrinos is their size. They are so tiny that they pass through matter without touching or affecting protons, neutrons, electrons and any other material objects.

In a startling announcement made in September 2011, a group of European physicists announced that in a very precise experiment, neutrinos travelled slightly faster than the speed of light. If this proves to be true, then many theories about the universe might be seriously affected. If proven true, neutrinos might be able to "take a short-cut through space, through extra dimensions." (See Overbye in the bibliography.)

In the future, neutrino telescopes may probe the universe with much greater detail than today's observational tools.

What is an atom?

The atom is a major synergistic structure that combines protons, neutrons and electrons. Most of the 92 natural atoms (elements) are forged in the dying furnaces of very large stars that have exploded as supernovas. There exist hundreds of varieties of atoms, depending upon the additions of extra neutrons, the loss or gain of electrons, and the myriad number of specialized architectures that form from varying structural combinations.

Like much of existence, the elements tend to organize themselves into hierarchies, such as periodic columns and rows.

The more complex the atom, the more protons and neutrons, the greater number of electron orbitals exist; and in many cases the orbitals permit openings for other elements to connect to one another by electron bonding to form an incalculable number of molecules.

The carbon atom is the foundation of life forms. It has four electrons available to make bonds with other elements, especially hydrogen, for strong, very long chains of molecules which are required for life forms. Essential atoms for life include carbon, hydrogen, nitrogen, oxygen, phosphorous and sulfur.

ISOTOPES

As already discussed, the addition of one or more neutrons to an atom creates a new, synergistic form with different properties from the parent atom.

ALLOTROPES

Some elements can exist in different molecular forms because of different structural modifications of the atom to permit them to be bonded in different ways. The classic examples are two of the allotropes of carbon—diamond and graphite. The carbon atoms are formed in different arrangements and thus these other synergistic forms are created.

MOLECULES

Two or more atoms that form an electrically neutral group and are held together by chemical bonding constitutes a molecule, from such simple arrangements, like water or salt, to the vast array of organic chemicals that give rise to living things, like the spectacular DNA molecule.

ISOMERS

Just as certain proteins in the living cell change function when they are arranged in different patterns or in a specific architecture, so too do chemical compounds that have the same chemical formula but act differently when arranged in a different geometric fashion.

Once again, we see that structure or form instigates process.

What is a star?

Our Sun is a star and is more massive than 90% of the stars in the universe. Its thermonuclear reactions power the solar system by emitting light, heat and many other radiations of the electromagnetic spectrum. In addition, the Sun produces cosmic rays which include particles like protons and hydrogen nuclei.

The Sun was created from a rotating disk of gas about five billon years ago and is estimated to have another five billion before its death. In its beginning stages, it was a protostar for about 250,000 years until temperatures reached 15 million degrees on the Kelvin scale, which is when thermonuclear reactions began.

As for intelligent life, the formation of stars is the single most important physical event to have occurred in the universe. Without exploding stars there would be no heavy atoms, like carbon, oxygen and iron, to allow for the creation of solar systems and planets.

Stars evolve from immensely large molecular clouds, mostly hydrogen, which form clumps thanks to the gravitational attraction and pressure which eventually fuse the hydrogen atoms in the core of the new star, thus creating nuclear reactions.

When a large dying star explodes into a supernova, there can be up to 92 elements formed in the explosion, and these elements eventually create material for the creation of other stars (like the Sun) as well as the seeding of planets (like the Earth).

No two stars are the same, as each one becomes a different type depending upon the size, brightness, surface temperature, stages of development and decline, and many other factors.

But the one event almost all have in common is the conversion of hydrogen into helium and heavier elements. This conversion is the source of the immense radiation given off by stars for most of their existence. The really giant stars do not exist long enough for life to emerge in their systems. As stars age and burn out of hydrogen, they undergo many different types of transformations that lead them eventually to become novas or supernovas, black dwarfs, neutron stars or black holes.

As for a black hole, the gravitational force is so immense that light and all other matter are sucked into it. Some have suggested that black holes may be the source for big bangs and new universes. There is evidence to show that radiation can escape from black holes, so they may not be as black as we once thought. Many cosmologists believe that the galaxies throughout the universe are in part powered by enormous black holes in their centers.

The Milky Way Galaxy where we reside contains about 200 to 400 billion stars, some of which are black holes, including the super massive black hole in the center of the galaxy where astounding cosmic pyrotechnics occur. Intelligent life at the center of the galaxy is quite unlikely.

Stars in galaxies form different structures, like open or globular clusters. An open cluster emerges from the formation of thousands of stars from a gigantic molecular cloud. Open clusters usually contain spiral arms like the Orion arm near which we find our solar system located.

A closed cluster contains thousands to millions of old stars bound together by gravitation.

Our solar system contains one star, but in more than two-thirds of the universe, binary or multiple star systems exist and because of their unusual orbits many spectacular events occur, such as pulsating x-ray sources, violent stellar explosions and immense and powerful outflowing jets of particles and gas.

Fortunately for us there is only a single star in our solar system, as multiple star systems would most likely create havoc and not allow intelligent life forms to exist in most instances. Also, some stars are variable in their brightness and would not be amenable to fostering intelligent life forms.

The solar system

As the Sun was created, it formed a condensation of atoms and molecules, such as silicon and iron compounds, as well as carbon dioxide, ice and water, some of which pre-existed in the original cloud that eventually formed the star. These compounds were affected by the solar wind, gravity and collisions among each other; the larger these bodies became, the more they attracted other materials, eventually to form planets, moons, and all other non-stellar bodies.

The inner planets of Mercury, Venus, Earth and Mars are made of solid materials, while the outer planets are the gas giants (Jupiter and Saturn) with small cores and huge amounts of hydrogen and other gases. The farthest planets, Uranus and Neptune, are usually called the frozen giants.

The diameter of the entire solar system is about 177 billion miles. To indicate how isolated the solar system is, the distance from our Sun to its closest star, Proxima Centauri, is about 25 trillion miles. These distances are staggering!

Chance seems to have played an enormous part in the creation of our solar system (or any solar system for that matter), as the great cloud of dust and gas that formed the Sun may have been influenced by shock waves from a nearby supernova, magnetic fields, the size, mass and spin of the proto-star and other forces. In actuality, when stars are born, many times they are usually part of a few to several dozen stars forming from the same giant gas cloud.

As the star forms, because of gravity, gas pressure and its spin, it tends to flatten out like a pancake with a bulge in the middle. Changes in temperature and other factors may have led to the formation of proto-planets within the spinning star which were expelled into space. (This is one of several possible theories.)

The first arrangement of the planets in their orbits around the new Sun could have taken about 100 million years. It is estimated that 3.6 billion years ago the solar system and its planets and moons settled into their present, predictable orbits.

The following provides a few facts about the planets:

Mercury is located about 36 million miles from the Sun. It revolves around the Sun in 88 days. Its period of rotation is 58.7 days. It is a little over 3,000 miles in diameter. It appears as if the planet's poles are covered in ice.

Venus is located about 67 million miles from the Sun. It revolves around the Sun in about 225 days. Its period of rotation (which is retrograde, meaning it spins in the opposite direction of the other planets) is 243 days. Its diameter is about 7500 miles. Venus is enshrouded by a very thick atmosphere (including clouds of sulfuric acid) making observations of the surface possible only by special telescopes with infrared or ultraviolet filters.

Earth is located 93 million miles from the Sun. It revolves around the Sun in a little over 365 days. Its period of rotation is 24 hours. Its diameter is about 7900 miles. The Earth's only satellite is the Moon, the fifth largest moon in the solar system. Its rotation is synchronous with the Earth's, so that it always shows the same side to the Earth. The Moon has very important influences upon the Earth's rotation speed, the tilt of its axis, the tides, as well as the biosphere. The Moon also may contain large amounts of water ice beneath its surface.

Mars is located about 142 million miles from the Sun. It revolves around the Sun in about 687 days. Its period of rotation is 24.6 hours. Its diameter is about 4200 miles. It has two very small satellites. Mars may have ice beneath its polar caps and water and mud during its spring and summer seasons in some locations.

Jupiter is located about 484 million miles from the Sun. It revolves around the Sun in 11.8 years. Its period of rotation is about 9.8 hours. Its diameter is about 89,000 miles, by far the largest body in the solar system and could have become a star when the solar system first formed if conditions were right. There are about 63 moons of Jupiter. Four of these moons are among the most massive satellites in the solar system. They are Io, Europa, Ganymede (the largest moon in the solar system) and Callisto. Io is the most geologically active body in the solar system with some 400 active volcanoes. Europa is thought to have a substantial amount of water surrounding its mantle and may harbor some form of life. Jupiter's enormous gravitational field is largely responsible for preventing asteroids from impacting upon the Earth.

Saturn is located some 887 million miles from the Sun. It revolves around the Sun in a little over 29 years. Its period of rotation is about 10 hours. Its diameter is about 75,000 miles. There are over 60 moons of Saturn, especially Titan, the second largest moon in the solar system, which contains a dense atmosphere and possible signs of water.

Uranus is located some 1.8 billion miles from the Sun. It revolves around the Sun every 84 years. Its period of rotation is 18 hours (also retrograde). Its diameter is about 33,000 miles. It has about 27 moons.

Neptune, the farthest planet, is located about 2.8 billion miles from the Sun. It revolves around the Sun every 165 years. Its period of rotation is 19 hours. Its diameter is about 30,000 miles. There are about 13 moons of Neptune, including Triton which has a thin atmosphere with clouds and hazes. It is one of the largest moons in the solar system.

It seems quite obvious that no significant design was apparent in the formation of the solar system. Deterministic principles, such as gravity, spin, rotation and chance occurrences are the operating factors in the formation of our solar system, as they would be in all other such systems.

But it must be made absolutely clear that the totality of the solar system—all its bodies revolving and spinning around the Sun, as well as moons revolving around planets, and the incalculable forces at play—is a synergistic whole, greater than each component and acting as a unified field which changes constantly in its dynamical relationships with one another and its position to the galaxy and beyond.

Adequate wisdom can never rule out celestial influence on the geosphere, hydrosphere, atmosphere and biosphere.

What is a galaxy?

Galaxies are distinct and vast collection of stars, interstellar gases, dark matter, nuclei (central bulges or cores), black holes and other exotic objects. There may be more than 150 billion galaxies in the visible universe, ranging in star population from several million to 100 trillion, all of which rotate around the galactic bulge or nucleus which is surmised to consist of a super massive black hole. The WOW! factor certainly applies to these massive galactic forms and processes which punctuate the entire universe.

They are influenced by synergy as they form hierarchical structures known as groups, clusters, superclusters and gigantic structures known as walls, sheets or filaments surrounding very large voids.

Although they come in many varied shapes, there are three major classifications, including galaxies like our own which form a spiral shape; those that form an elliptical shape; and then many irregular shapes which result often from the gravitational pull from other galaxies or the cannibalization of one galaxy by another. Star formation in some galaxies may continue for perhaps another 100 billion years!

Little is known about galaxy formation, but they may have formed out of super massively large molecular clouds some 500 million years after the Big Bang. For each galaxy, a wide variety of molecular clumps are spread out over thousands of light years (in our galaxy about 100,000 light years in length) where star formation takes place.

There is much uncertainty about galaxy formation and whether stars or galaxies formed first. If we subscribe to the process of synergy, we would favor the idea that stars came first and then began to form galaxies. (If we were to subscribe to the preformation theory, then we might suggest that the very structure called a galaxy is built into the fabric of the universe. Stars are simply the forms and processes leading to an already determined galactic form. This notion is of course quite speculative, but we cannot rule it out.)

GROUPS OF GALAXIES

Galaxies near to one another form groups that are gravitationally linked to one another. The Milky Way has about 30 such galaxies "held" together by the pull of the Andromeda Galaxy and the Milky Way. They are called the Local Group.

CLUSTERS OF GALAXIES

Clusters of galaxies may contain from 50 to 1,000 galaxies held together by gravitation as well as the mysterious dark matter about which we know literally nothing! The Virgo cluster contains the local group of galaxies, including the Milky Way.

SUPERCLUSTERS OF GLAXIES

Galactic clusters form superclusters. The Milky Way and the local group and the Virgo cluster are part of the Virgo supercluster which contains about 100 galaxy groups and galaxy clusters about 110 million lights years in diameter.

FILAMENTS OR GREAT WALLS OF GALAXIES

These are the largest structures in the universe. There are about 10 of these massive forms, the largest being the Sloan Great Wall, which is an amazing 1.37 billion light years in diameter and occupying about $1/60^{th}$ of the observable universe. Filaments or walls are separated by immense voids containing little or no matter.

The celestial frames of reference

Through what prism do we view the universe? In ancient and medieval times, the Earth was thought to be the center of the universe, and that everything revolved around it. This was a form of the geocentric frame of reference, which has been expanded to include the heliocentric frame that shows how we are part of a solar system in which all things revolve around the Sun. An even larger frame is the galactic reference, showing how our Sun is just one of several hundred billion stars throughout the Milky Way. Then there is the local group of galaxies to which we belong, another frame. Then we have a galactic cluster frame (the Virgo cluster) affecting our local group of galaxies, and so on.

There is misinformation about the use of the geocentric frame of reference. Admittedly the Earth is not the center of the world, but it is the recipient of celestial influences from the Moon, Sun, and other bodies in the solar system and the universe. As the Earth spins on its axis daily, the entire universe appears to revolve around it. The spin of the Earth moves the horizon clockwise about one degree every four minutes, presenting a continuously moving reference point, allowing for a 360 degree frame of reference each day. The yearly slightly elliptical revolution of the Earth around the Sun forms a great circle known as the ecliptic, and since most other planets and the Moon follow this path, partial or full eclipses can occur from time to time. The Earth's revolution, as well as its axis tilted about 23.5 degrees, create our seasons.

The geocentric view of the solar system and the universe is as valid a view as the heliocentric frame of reference.

ADEQUATE WISDOM

PART FOUR:

BIOLOGICAL & HUMAN EXISTENCE

In Part Four, there are brief descriptions of the major components of biological existence, including the human cell, tissue, organ and organ systems. Special attention is paid to the human brain and its major components. Examples of chance and determinism and their effects upon the human body are presented by pointing out how dysfunctions in the chromosomes can bring about illness and disease. Finally, there is a discussion of human personality.

What is life?

One of the most spectacular synergistic events of existence and the universe has been the formation of life forms on Earth beginning about 3.8 billion years ago. Whether the organic "seeds" that eventually created life originated on Earth or were transmitted here from meteorites, comets, asteroids or planets is still hotly debated. There is also the possibility that life sprang from the Earth's mantle, where forms of life existed without the need for oxygen. In fact, it has been suggested that there is more life mass and water beneath the surface of the Earth that above it!

Upon the advent of photosynthesis and the production of oxygen, and millions of years of trial and error, molecules in the form of long chain polymers developed the miraculous ability to form living cells which divided and then differentiated themselves into tissues and organs, especially the brain, allowing us to speak and write about this and all other phenomena in the world.

The majority of animal structures of the world first came into existence about 500 million years ago, so that billions of years of evolution led to their creation.

Biological evolution

There are about nine million species of life forms now existing on Earth, each following the dictates of their specific DNA master species code. Throughout biological history, species have evolved and then died off. Current thought claims that evolution results from chance mutations in genetic materials as well as natural selection, which means that evolution favors the species whose DNA changes allow it to survive and adapt in a robust manner alongside other species, climates and environments. Creatures evolve differently in different locations

There has been an ongoing debate about environmental influence and the unfolding of a zygote (fertilized egg) into a finished organism. Biologists use an absolute rule that claims that DNA (except for mutations) is a one-way path and cannot be influenced by external forces. But just as important is the claim that the environment can indeed influence the growth of an organism, acting not on the genetic codes but rather on the organ systems and organism and its maturation process.

DNA directs the formation of proteins which then assemble cells, then tissues, and then organ systems. But it must be recognized that organs and organisms are new wholes and can be influenced by the environment. According to synergy, the organs are greater forms and different from the cells and tissues that created them. Thus the organism, you and I, are not totally beholden to DNA, but instead become a new dynamic whole, greater than our DNA and organ systems and affected substantially by the environment and circumstance.

How life began is often explained in the following fashion. Over the course of several hundred million years after Earth cooled, a group of molecules formed the RNA polymer which contained enzymes able to duplicate itself and eventually form a living cell. The cell somehow grew to include several key organelles along with a master DNA molecule in the nucleus of the cell. Over the course of billions of years, evolution (comprising natural selection, genetic drift, variation, recombination and mutation) produced millions of different master DNA codes creating millions of specific species, including humankind. Some species become extinct and new species are formed. The odds of complex cells forming out of inert molecules are overwhelmingly slim, but yet here we are, which serves as the reason that many believe in some form of design.

Another viewpoint suggests that life emerged from the collection of molecules in the primordial soup as the result of a bridge between chaos and order—a kind of phase transition that creates life forms. This compares somewhat to the "acorn effect," which suggests that the essence of life rests within the chemical mixtures. Some kind of preformation principle and vital force are close relatives to this viewpoint.

Finally, the view that God has created life (as well as the universe itself) is held by the vast majority of human beings, partly because it is easier to grasp that a Maker exists and creates all the wondrous phenomena in the world; and also partly because this viewpoint provides solace to those who believe that without a God there would be no comfort or purpose in one's lifetime. However, purpose can be found in living a life without a Maker as long as each person or institution subscribes to the promotion of pleasure and responsibility and strength and compassion, using the guidelines of common sense and rationality.

The hierarchical arrangements of life forms

The living things on our planet are arranged in groupings as follows:

> Species (and sub-species)
> Genus
> Family
> Order
> Class
> Phylum
> Kingdom
> Domain
> Biosphere

Species are organisms capable of interbreeding and producing fertile offspring.

Genus is a sub-division of a family and usually contains more than one species.

Family represents those organisms evolving from the same ancestors.

Order indicates specific branches of life, such as primates (humans, apes, etc.).

Class represents a specific branch, such as mammilla (all mammals).

Phylum represents life forms with similar structure and an evolutionary relatedness.

Kingdom represents the major division of living things and is often classified as animal, plant, fungi, protista (one-celled forms), archaea (microbes) and bacteria.

Domain represents the three basic forms of life, depending upon the cellular structure, and are archaea (single cells), bacteria and finally eukarya (complex, multicellular organisms with highly developed nuclei).

The biosphere encompasses all living things.

The human cell

The living cell is the building block for the human organism, as well as many life forms on Earth, starting as the fertilized cell (zygote). Parental sex cells recombine in a synergy that initiates the development of a unique human being. Through a series of cell divisions, the new cells begin to take on special roles by forming tissue systems and organ systems until finally the birth of a new human occurs. It is estimated that each human has about 100 trillion cells, in addition to countless numbers of other organisms, primarily bacteria that help their human host function properly.

The one cell that creates an entire organism reminds us of the acorn, where everything necessary to build a beautiful oak tree resides inside one seed.

The instruction manual for human development, as for most other living things including oak trees, resides in the DNA material inside the primal cell. Determinism and chance bring about an incalculable number of forms and processes that allow cells to divide and mysteriously form all the necessary parts for the whole human.

Once again, we find a series of parts inside each cell that form into a synergistic structure with specific functions while also being part of the entire whole.

The major cell components are:

CYTOPLASM
The thick soup of mostly water that holds all the cell components together and allows for cell movement and electrical conductivity.

CELL MEMBRANE

This semi-permeable envelope provides the "skin" around each cell that regulates which substances can enter or leave the cell. The membrane also contains receptors that can detect signals from other cells.

CELL NUCLEUS

Protected by a double membrane, the nucleus is the information and administrative center of the cell, containing most of the cell's genetic materials in the form of long strands of chromosomes and DNA which contain genes that provide the codes for the manufacture of proteins (amino acids). Pores in the nuclear membrane allow various chemicals to enter and exit the nucleus to allow thousands of cellular transactions to occur.

RIBOSOME

This cell component translates the genetic instruction into proteins (amino acids) which form globular structures that increase or decrease chemical activities in the cell.

LYSOSOME

This part of the cell absorbs and eliminates waste products and debris, including microbes and dead cells.

MITOCHONDRIA

These organelles provide most of the energy for each cell, producing a compound called ATP (adenosine triphosphate). Since they also include DNA, it is believed that the original cells somehow captured these organic units and incorporated them into the cell structure. Mitochondria are also involved in cell growth and cell death, communication among cellular components and other activities.

GOLGI COMPLEX

This cell component is involved with modifying, sorting and packaging of large molecules that the cell secretes to other parts of the tissue and organs, as well as for use by the cell itself. It creates the lysosomes and transports fats and other materials throughout the cell.

ENDOPLASMIC RETICULUM (ER)

The smooth and rough versions of these organelles act as the transportation system throughout the cell, as well as taking part in the folding of certain proteins. Much like FEDEX or UPS, the ER assigns address codes when sending these proteins throughout the cell so that they arrive where they are needed. They also assist in drug detoxification.

When we consider how tiny a cell actually is and how many thousands of transactions occur continuously, we really have no words fully capable of expressing our awe and wonderment for that which permeates all living things!

The key organic compounds

For all advanced life forms, the DNA molecule (comprised of nucleic acids) contains all the necessary information for constructing an organism. It is organized in varying number of chromosomes for different species and contains genes that code for the building of amino acids, which through the RNA molecule code for the production of proteins that mostly present themselves as three-dimensional, globular shapes, each one of which has specific duties to perform in the cell. Some of these proteins are enzymes which speed up or slow down cellular activities. The three-dimensional architecture of these proteins determines their function.

Two other organic compounds are carbohydrates (sugars) and lipids (fats and steroids).

DNA represents the design feature of evolution, locked into place by determinism, meaning, with an important exception, that the cells and organs of advanced life forms are faithfully reproduced each and every time. We know that the exception is what creates diversity as chance events can alter DNA and make slight or significant changes leading to new adaptations and occasionally new species.

The key atoms comprising DNA are hydrogen, nitrogen, oxygen, carbon and **phosphorus**.

The key atoms comprising amino acids are hydrogen, nitrogen, oxygen, carbon, and **sulfur**.

Embryonic stem cell

Organs and organ systems unfold from the embryonic stem cell, which has the mysterious ability to form all the parts of the human. The stem cell comprises three layers, the ectoderm, endoderm and mesoderm, and from these three layers emerge the tissues, organs, organ systems and, finally, the complete organism.

The DNA's on and off switches direct the assemblage of these parts.

The ectoderm forms the central nervous system, the skin, as well as other parts such as the lining of the mouth, nostrils, and anus.

The endoderm layers develop into the urinary system, endocrine glands, respiratory tract, gastrointestinal tract, urinary system and auditory system.

The mesoderm layer unfolds to produce bone, muscle, connective tissue and the middle layer of the skin.

Human tissue systems

There are four types of tissue systems as follow:

1. Cells that are packed tightly together to form sheets to serve as linings for different parts of the body, which form membranes to protect and keep organs separate from one another.

2. Connective tissues provide support and structure for the body, such as tendons, ligaments, cartilage and fat tissue.

3. Tissues that allow muscles to contract.

4. Tissues that form neurons and glial cells. (Glial cells are electrically neutral and support the neurons by providing oxygen and nutrients, while destroying pathogens and removing dead neuron cells.)

Human organ systems

Usually working together like a well oiled machine, the human organs exhibit yet another example of individualism and synergy, since each one has its specific functions while at the same time cooperating together to form a synergistic whole.

These organ systems include:

CENTRAL NERVOUS SYSTEM (CNS)
This is the executive center of the body that receives and processes information and regulates body functions. It includes the brain, spinal cord and peripheral nerves. It is responsible for consciousness, thought and creativity; as well as countless other body functions.

CIRCULATORY SYSTEM
This pumps and channels blood throughout the body and includes the heart, blood and blood vessels.

ENDOCRINE SYSTEM
This is a collection of glands that secrete hormones into the bloodstream that regulate many body activities. Such glands include the adrenals, the thyroid, pituitary, hypothalamus, pineal and others.

DIGESTIVE SYSTEM
This system deals with the digestion and processing of food, and encompasses the salivary glands, esophagus, stomach, liver, gallbladder, pancreas, intestines, rectum and anus.

IMMUNE SYSTEM

This system protects the body by fighting off disease from bacteria, microbes, viruses; and includes white blood cells, tonsils, adenoids and spleen.

MUSCULOSKELTAL SYSTEM

This includes the muscles and the skeleton for structure and movement, including bones, cartilage, tendons and ligaments.

REPRODUCTIVE SYSTEM

These include the sexual organs. For males, they include the testicles, penis, prostate, seminal vesicles and vas deferens. For females they include the ovaries, fallopian tubes, uterus, vagina and breasts.

URINARY SYSTEM

This includes the bladder, kidneys and urethra.

RESPIRATORY SYSTEM

This includes the pharynx, larynx, trachea, bronchi, lungs and diaphragm.

INTEGUMENTARY SYSTEM

This includes the skin, hair and nails.

The human brain and the neuron cell complexes

Of course we are fascinated by the entire central nervous system that supervises and controls most body functions and represents a synergy unsurpassed by any other organ system. But the key feature is the neocortex, a sheath of six layers covering the cerebral hemispheres. Along with its many duties, the neocortex acts as the center for thought, reasoning and language, aided of course by other brain centers all working as interconnected forms. The neocortex is the center for free will, human design, volition and creativity.

THE ABSOLUTELY MARVELOUS NEURON

The basic neuron is a cell like any other in the body with a nucleus and many organelles carrying out the incalculable processes to remain viable and functional. But, the basic neuron also serves as a transmitter and receiver of electrical and chemical signals that turn on or off other cells and chemical reactions in the brain.

The average neuron has many different branches, called dendrites, which look like the branches of a tree and which extend outward to make chemical or electrical connections to other dendrite trees and chemical mixtures nearby. These neurons are called the grey matter of the brain. There are hundreds of specific neuron types with different branch patterns performing specialized tasks.

Virtually every neuron also contains an axon extension which sends and receives communications from other neurons and the chemical flux surrounding all neurons. The axon, unlike the dendrites, can travel great distances, up to many feet, and serves as the white matter of the brain, forming bundles that deliver messages to neuron complexes within and between the two brain hemispheres.

Axons are the great communicators in the brain and keep other neuron complexes aware of what's happening at all times.

Brain components

The three components of the human brain are the:

Cerebral cortex—the command center for higher neuron functions. It is a thin mantle of neurons covering the surface of each cerebral hemisphere. The cerebral cortex is crumpled and folded, forming numerous convolutions and crevices and is responsible for the processes of thought, perception, volition and memory. It also acts as the seat of advanced motor function, social abilities, language, speculation and problem solving.

Cerebellum—the center for coordinating our senses, movements, balance, muscle tone and much more.

Brain stem—connected to the spinal cord, the brain stem manages messages going between the brain and the rest of the body, and also controls basic body functions such as breathing, swallowing, heart rate and blood pressure. The brain stem also controls consciousness and determines whether one is awake or asleep.

BRAIN HEMISPHERES

The human brain is divided into two distinct hemispheres connected by white matter, which is a large number of axons (bundled nerve cells) permitting communications between the two spheres.

While it is assumed that the hemispheres concentrate on specific functions, many times an injury to one sphere's function can be compensated by the other sphere. There is definitely the sense that the entire brain is an exquisite synergistic whole of trillions of electrochemical interactions.

However, each hemisphere shows dominance for certain characteristics, such as language skills exhibited by the left hemisphere in a vast majority of right-handed individuals. Language, science and math skills, reasoning, grammar and vocabulary are mostly managed by the left hemisphere; whereas certain speech intonations, creative, artistic, visual and auditory (music) functions and skills seem to show right hemisphere dominance.

LOBES OF THE BRAIN (The Four Divisions)

Frontal lobe—responsible for speech and language, decision making, planning, abstract thought, volition, creativity; as well as skilled muscle controls and movements.

Parietal lobe—responsible for relaying all sensory stimuli to other locations in the cortex.

Temporal lobe—concerns itself with hearing and memory skills.

Occipital lobe—the visual processing center of the brain.

THE LIMBIC SYSTEM

Comprised of many different neuron structures, the limbic system is linked to the cerebral cortex and deals with basic human drives, temperature control, hormone production, emotions, longterm memory, sense of smell and much more.

THE AMYGDALA

Part of the limbic system, this is the emotional center of the brain and may serve as one source of personality disorders, since atypical neuron firings may produce excessive hate, anger, fear and other emotions as well. Damage to this region could eliminate emotional reactions to everyday experiences. Recent studies have shown that those living in urban areas are subject to more stress in this region of the brain than those living in less dense, less chaotic geographic areas. (Which certainly makes sense.)

THE HIPPOCAMPUS
This brain part largely deals with memory and the recall of events in the lives of a human being. Damage to this region is responsible for amnesia. This appears be the only brain region where new neurons are generated. In all other parts of the brain, when a neuron expires, a new one does not replace it.

THE CINGULATE GYRUS
This region impacts upon the impulsive nature of a human, with damage to it perhaps leading to attention deficit disorders and emotional outbursts.

CORPUS CALLOSUM
Several hundred million axons (white nerve fiber) connect both brain hemispheres and permit the integration of all brain activities into a synergistic whole.

A standing ovation for the brain

Whether the brain evolved by chance and determinism or whether it was somehow programmed really does not matter when marveling at its stunning success. Directing, coordinating, regulating and analyzing the organ systems of our bodies and managing our five senses on a 24/7 basis, the brain deserves a tip of the societal hat and stentorian praise for its wondrousness.

We know that nothing works in a vacuum, so not only do we congratulate our brains but we also need to credit all the other synergistic couplings that helped produce the brain.

The evolutionary chain of forms and processes led to the advent of single cell organisms which led to multiple cell organisms which led to complex cells and complex molecules leading to marine life, the greening of land and the explosion of mammalian populations; the latter with burgeoning brains, much of which was due to the need for enhanced smell. Most of the brain growth in early mammals occurred in the olfactory bulb. But as times got trickier and more dangerous for mammals to survive, a newer, smarter brain evolved, equipped with a neocortex that propelled Homo sapiens into the forefront of cognition and planetary command.

Free will, chance and determinism enable us to use the brain for good or evil. Therefore it becomes the duty of compassionate citizens to make certain that policies are enacted and enforced that promote the well-being of humanity.

Human chromosome malfunctions

Each of the 23 pairs of human chromosomes can have one or more defects that can seriously affect the life of a human being. In addition to inherited diseases or tendencies, there are many ways in which a person's genotype becomes flawed. Chromosomes can be affected by mutations, a loss of material, extra copies, a reverse in the direction of a chromosome or broken parts connecting to other chromosomes.

The largest human chromosome, number 1, contains over 4200 genes, and mistakes or mutations can cause or contribute to 890 known diseases, such as deafness, Alzheimer's disease, glaucoma and breast cancer.

Chromosome 2 malfunctions can contribute to autism, ALS, hypothyroidism, spastic paralysis, colorectal cancer, juvenile diabetes and more.

Chromosome 3 malfunctions can contribute to autism; cancers of the breast, colon, ovaries, lung and pancreas; cataracts; brain disorders; deafness; diabetes; heart blockage; lymphomas; night blindness; short stature; leukemia; and more.

Chromosome 4 malfunctions can contribute to bladder cancer, leukemia, muscular dystrophy, hemophilia, Huntingdon's disease, deafness, Parkinson's disease, kidney disease and much more.

Chromosome 5 malfunctions can contribute to impaired development of the nervous system; cat-like cries of children; diseases of the eye; nicotine dependency; Parkinson's disease; spinal muscular dystrophy and more.

Chromosome 6 malfunctions can contribute to blood disorders, deafness, kidney disease, Parkinson's disease, rheumatoid arthritis, diabetes, epilepsy and more.

Chromosome 7 malfunctions can contribute to brain malformation; motor and sensory disorders; cystic fibrosis; muscular atrophy; blood disorders; colon cancer; childhood diabetes; deafness; schizophrenia; color blindness and more.

Chromosome 8 malfunctions can contribute to cleft lip and palate; thyroid malfunctions; bone spurs and lumps; schizophrenia and more.

Chromosome 9 malfunctions can contribute to defects in replicating cells; neurological dysfunctions; skin and endocrine diseases; deafness; renal and bladder stones and more.

Chromosome 10 malfunctions can contribute to malformations of the skull, face, hands and feet; disorders of the nervous system; enlargement of the colon; skin discoloration; and more.

Chromosome 11 malfunctions can contribute to autism; breast and bladder cancer; Mediterranean fever; leukemia; liver and spleen malfunctions; sickle cell anemia; deafness, blindness and more.

Chromosome 12 malfunctions can contribute to underdeveloped limbs; connective tissue diseases; severe problems of the eyes; narcolepsy; Parkinson's disease; bone growth disease and more.

Chromosome 13 malfunctions can contribute to bladder and breast cancer; colon disorders; deafness; sudden death; eye disease; and more.

Chromosome 14 malfunctions can contribute to blood and lung dysfunctions; Alzheimer's disease; thyroid disorders; blood cancer; deafness and more.

Chromosome 15 malfunctions can contribute to problems of growth and development of the body; breast cancer; Tay-Sachs disease; metabolism dysfunctions; and more.

Chromosome 16 malfunctions can contribute to miscarriage, Crohn's disease, blood disorders and more.

Chromosome 17 malfunctions can contribute to neurological degeneration; bladder and breast cancer; connective tissue disorders; central nervous system malfunctions; brittle bone disease; mental retardation and more.

Chromosome 18 malfunctions can contribute to abnormal blood vessels; speech disorders; malfunctions of the sex chromosomes; mental retardation and more.

Chromosome 19 malfunctions can contribute to Alzheimer's disease; strokes; thyroid disorders; migraines; body weakness; brain damage; muscular dystrophy; brain tumors and more.

Chromosome 20 malfunctions can contribute to skin and joint diseases; small intestine disorders; juvenile diabetes; brain disorders; deafness and more.

Chromosome 21 malfunctions can contribute to Alzheimer's disease; ALS; Down syndrome; metabolic disorders; heart disorders; deafness; and more.

Chromosome 22 malfunctions can contribute to ALS; breast cancer; tumors; central nervous system malfunctions; learning difficulties; blood disorders; schizophrenia and more.

Chromosome 23 (X for females, Y for males) determines the gender of each human. Some of the disorders from malfunctions in the X or Y chromosome can produce low levels of testosterone, learning disabilities, low IQ, infertility, feminine attributes in males, masculine attributes in females and more.

Chance and determinism (heredity) bring about chromosomal dysfunctions that can impact greatly upon the lives of many unfortunate human beings, which often lead some cynical people to claim that "life is a crapshoot."

What is personality?

Personality is a collection of traits that have recurred throughout human history—no matter if it's thousands of years ago or today. There exist a large number of these recurring traits including such examples as humor, greed, love, generosity, hoarding, extroversion, introversion, pessimism, optimism, amiability, meekness, jealousy and the like.

It is quite important to point out that much of how we present ourselves to others and how we appreciate life rests upon the personality complex within each human being.

Personality traits (and emotions) sometimes appear as part of a continuum, so that traits or emotions like happy and sad form two ends of a spectrum. During the course of a lifetime, an individual will invariably find himself or herself in a happy, neutral or sad mode. Other continuums include optimism—pessimism, love—hate, greed—and generosity. Few people exist in the middle of a continuum, so that there are those who, for example, are sadder than others or more mischievous than others.

Now what could account for these traits? Are they part of the genetic package and the brain chemistry? Are they learned? Does natural selection have any role in their development and presence in humans?

We actually do not know. *The fact that they are persistent and recurring phenomena suggests that other recurring forms and processes may be creating, influencing or sustaining these traits historically throughout humankind.*

It seems as if they are not learned. The reason for this is that children do not tend to exhibit the same traits of their parents or siblings. A child raised by generous and philanthropic parents can be quite the opposite, that is, greedy and self-centered; good humored parents can often see their child's personality develop as cool and detached, without a sense of humor. Siblings raised by the same parents in the same household exhibit different traits. All parents recognize the personality of a child fairly early in its development, before significant socialization takes place. Children do not acquire personality traits from their friends nor from any other social structures.

Genetic chemistry might account for some personality traits. This would entail a range of specific genes that code for personality and would mean that a pot-luck selection yields specific traits for each person, which seems unlikely.

There are reward centers in the brain, pleasure centers if you will, that certainly influence personal behavior and character, but research is limited at this time. There are several important brain complexes that process emotions, but we really do not know whether the traits exist prior to or after the neurons trigger chemical and electrical impulses. Some traits, like love, fear and hate, probably are not personality traits, but rather emotions, about which we also know little.

Natural selection does not seem to be a factor in personality development. If it were, personality traits inimical to the species would have been weeded out a long time ago.

One other factor, quite a tremendous taboo in science and religion, is the influence of the cosmos, mainly the state of the solar system at any given moment in time. Quantum mechanics has now shown that action at a distance is a viable phenomenon, so that the processes and forms of the solar system field, which vary second by second, may indeed influence the biosphere. The Earth's daily rotation, yearly revolution and longterm precessional movement, as well as the constantly changing angles among the planets, moons and Sun, create unique fields at any given time. In fact, the planets account for 98.5% of the angular momentum (rate of spin) of the Sun!

We all acknowledge the influence of the Sun and Moon on weather, tides and other planetary phenomena. Is it possible that celestial influence also plays a part in human personality and society?

To bolster the idea of celestial influence, refer to some of the writings of physicist Brian Greene, whose 2004 book, entitled *The Fabric of the Cosmos*, discusses different aspects of quantum theory, including the notion that "something you do over here can be instantaneously linked to something happening over there, regardless of distance." He states that "research has confirmed that there can be an instantaneous bond between what happens at widely separated locations." (P. 11). Further, he states that "theoretical and experimental considerations strongly support the conclusion that the universe admits interconnections that are not local." (P.80). These comments suggest that more than gravity and electromagnetism influence our planet.

This stunning principle of quantum mechanics may convince the naysayers of celestial influence eventually to initiate some analytical research to be conducted in this area.

It is unfortunate that close-mindedness has prevented further study of celestial influence. One day, there may arise a melding of biology, psychology and celestial mechanics to permit further study. This new investigation would require ending the unjustified taboo created by science, social science and religion.

But no matter what factors are at play in creating the personality of each human, a mature adult comprises a collection of influences that result in a person's character, which is the compilation of traits, emotions, inclinations, proclivities, social and physical intercourse and aesthetics that make each one of us unique living beings.

Personality types

There are discernable types of personality and character traits that seem to express themselves in greater or lesser strength among the human population.

Such different types include those with a passion and exuberance for life. Those who are reserved and quiet. Those who appear emotional and moody. Those who are filled with lust and sensuality. Those who seem detached and reflective. Those who are greedy and who hoard possessions and are stingy towards others. Those who are jealous and subject to rage. Those who are cheerful and pleasant. Those who are vain. Those who are meek and frightened by the world at large. Those who show off their talents, their possessions and their appearance. Those who express confidence and success.

Those who are impatient, foolhardy or impulsive. Those who are witty and conversational. Those who are cunning. Those who are protective and sympathetic. Those who are faithful. Those who are dogmatic and resistant to change. Those who seem very critical of other people. Those who are charismatic. Those who exhibit no guile and are straightforward in their dealings with others. Those who are flighty. Those who change their minds often.

Those who are practical and patient. Those who are optimistic. Those who are pessimistic. Those who are innovative. Those who express a strong sense of independence and self reliance. Those who are perverse and seek no good for others. Those who are sensitive to others. Those who are unworldly and spiritual.

Each person can exhibit a variety of these traits. It seems very likely that recurring forms and processes bring them about and sustain their unique combinations in humans. The reasons why these specific traits exist are still uncertain, but the previous essay offers some suggestions.

The blend and clash of personalities

The forms and processes of human personality and character create endless varieties of social interactions among the population. One-on-one relationships include soul mates, best friends or worst enemies. Personalities in these relationships are either complementary or antagonistic. In larger groups, say a teacher in a classroom of students or an employer and his staff, there is an overriding personality that dictates the social interactions. In theory, the subservient personalities give way to the dominant one. In reality, there are many occasions where personalities clash with varied consequences, and thus the brash student may be punished or the contentious worker disciplined or fired.

Strong personalities usually include charismatic people whose presence and will power can dominate others. People with such dominance and ambition will often rise to the top of their profession in a commanding fashion. Often times those with strong personalities attribute their success to free will, while not considering the effects of chance, determinism and synergy.

Weak personalities are subservient and withdrawn. They often follow and do not take the lead in social intercourse. Those with strong personalities can often take advantage of weaker people and manipulate them to their advantage.

Personalities of all types do not simply appear out of the air. They are the result of the combination of forms and processes. They emerge early in life and eventually form unique characters during the defining years when social interactions begin to multiply.

Those with weak, moderate or strong personalities can exhibit a different type of character. A weak-willed individual can still have a good heart and a compassionate nature, while a strong personality's character can be rough and uncompassionate.

The interplay between personality and character makes for a robust range of human interactions that are influenced by chance, determinism, human design, free will and synergy.

The brain and unusual human behaviors

Studies of the brains of those with antisocial, criminal or personality disorders or marked mood swings point to a group of neurochemicals and hormones that may trigger unusual behavior when they are not in balance, meaning that too little or too much of each one or more is produced.

Some of these include:

> Dopamine
> Serotonin
> Beta-endorphins
> Norepinephrine (Adrenaline)
> Steroids

Some or all of these in an unfortunate combination can trigger mood swings, criminality, impulsivity and aggression, antisocial actions, psychosis, depression, anxiety, bipolar disorder, attention deficit disorder, irritability and substance abuse.

There are genetic and environmental factors at play here, but no one is absolutely certain just how these neurochemicals go out of whack. It is quite possible that, in addition to genetic malfunctions, environmental factors can create the stress that triggers them.

Nevertheless, once again we see how chance and determinism work. Individuals who exhibit such behavior cannot be blamed for their actions. And fortunately, some of these behaviors can be modified by pharmaceuticals.

Incidentally, the simple physical act of laughing can trigger the release of endorphins to alleviate pain and stress, as well as possibly promoting the overall health of an individual, reducing the need for pharmaceuticals.

Can we rise above circumstance?

We all know people who are grumpy, malcontent and cynical about their lives, their families and their jobs; whose traits and circumstances have not been favorable. There are those who are rude to others, pushy and arrogant and generally displeased more often than not.

The power of chance and determinism is not to be underestimated. When both of these modes of existence work against an individual, it takes much effort to maintain cheerfulness about life and to rise above the muck and transcend the miseries that many people suffer.

The smaller institutions, like families and peer groups, can help by way of positive intervention, using human design and free will to create ways to circumvent unfavorable processes. Even larger institutions could engage specialized personnel officers who seek out malcontented people and work with them to formulate policies to alleviate stress and uncertainty. (More often than not, disgruntled people are dismissed and left to their own devices.)

Perhaps all social institutions may one day construct policies that actually help unfortunate people make adjustments in their lives and demonstrate compassion for the discontented people of the world.

We know that pharmaceuticals which modify brain chemistry are in great use and demand, and do allow people in some cases to rise temporarily above their unfortunate psychological and physiological circumstances. These do have side effects, some mild, some severe.

We know that national and state institutions and foundations, assuming they have the funds, offer loans, grants and scholarships to people whose circumstances would have otherwise prevented them from attending schools of higher education. This is right thing to do. Social structures must always work to alleviate the physical, mental and economic agonies that people face.

What about people born into ignorant and prejudiced families; born in countries that are ruled by dictators who do not provide benefits for its citizens; born in sheer poverty? Circumstances can be cruel, and it would take a person with powerful traits and will power to rise above these conditions.

Then there are times of war when citizens and combatants are all subject to injury and death.

The sad conclusion to our original question is that at this point in our evolution most unfortunate or malcontented people cannot rise above difficult circumstances and must toil away in a harsh, boring or antagonistic environment.

Those who claim that each person is capable of pulling himself or herself up from the muck and accomplishing anything they want are not looking at the real world. These people are usually the rugged individuals who believe that what they accomplished in their lifetimes was completely the result of their own efforts, without any nod to chance and circumstance. They propagate pure individualism without any recognition of synergy, and are guilty of promoting a half-truth.

Of course some few can indeed shine through the mess they have been given, but for most of the poor souls ensnared by unfortunate instances of chance and determinism and an underachievement of free will, their lives are sad and disheartening.

Thus the longterm goal of any humanistic civilization is to create policies that soften the blow for its less fortunate citizens and eventually create equal opportunities and compassion for the entire human population.

PART FIVE:

TRENDS & OTHER MATTERS

Part Five provides an approximate timetable of existence, from the Big Bang through the beginnings of human civilization. There are brief essays about the quantum universe, relativity theory and the implications of infinity. We ask if there will be an end to the universe, and discuss the apparent trend toward synergy.

Cosmic and biological evolution:
An approximate and condensed timetable

THE PHYSICAL UNIVERSE

13.7 billion ago: the "birth" of our universe occurred called the Big Bang.

Nanoseconds after the Big Bang, a common temperature was reached.

Nanoseconds later a 100-fold expansion of space occurred in what is called the Inflationary period. Quantum fields and anti-gravity generated a repulsive energy that accounted for the immense expansion of space and the generation of particles. During the first three minutes, the intense radiation from quantum perturbations transformed the energy into particles, like quarks, neutrons, protons and electrons. At about this time, the first atoms were created, mostly hydrogen (about 75% of matter), helium (24%) and a trace of lithium.

About 370,000 years after the expansion of the universe, photons (light) broke free of the plasma of charged particles and began to stream throughout space and now resonate everywhere in the form of microwave background radiations, providing proof of the spectacular "birth" of our universe.

About five hundred million years after the Big Bang, the first stars and galaxies appeared, assembled by the tendency for matter to form clumps under the force of gravity and pressure to create thermonuclear reactions in the centers of rotating disks of mostly hydrogen gas. (The evolution of galaxies is one of the great puzzles to be unraveled. Nothing is really certain about their development.)

THE SOLAR SYSTEM

5 billion years ago: formation of the proto-Sun.

4.6 billion years ago: the Earth and solar system form.

4.5 billion years ago: a very large body was assumed to have crashed into the Earth and from this collision the Moon was created. But now new research suggests that initially the Earth had two moons which eventually collided into one another to form the current Moon.

4.1 billion years ago: the earth's crust cools and solidifies. An early atmosphere and oceans appear.

3.9 billion years ago: a shift in the orbits of the outer planets sent a surge of comets and asteroids into the inner solar system.

THE APPEARANCE OF LIVING THINGS

3.8 billion years ago: the first simple cells emerge.

3 billion years ago: the advent of photosynthesis and oxygen production.

2.5 billion years ago: first major ice age.

2 billion years ago: more complex cells emerge.

1 billion years ago: multicellular life forms appear.

800 million years ago: second major ice age lasting almost 200 million years.

600 million years ago: simple animals emerge.

580 million years ago: the ozone layer forms.

570 million years ago: ancestors of insects, arachnids and crustaceans appear.

525 million years ago: Cambrian explosion. The biological structure of complex animals was formed. A wide diversity of living things appeared.

500 million years ago: fish appear, along with early amphibians.

475 million years ago: the appearance of plant life on land.

400 million years ago: insects and seeds form.

360 million years ago: the rise of amphibians.

300 million years ago: reptiles emerge.

250 million years ago: a mass extinction of more than 90% of marine life occurs.

225 million years ago: mammals appear.

200 million years ago; the earliest dinosaurs appear.

150 million years ago: birds appear.

130 million years ago: flowers and bees appear.

80 million years ago: ants and termites appear.

65 million years ago: the dinosaurs die out, as well as about half of the animal species, probably from an enormous meteor impacting the Earth.

40 million years ago: modern butterflies and moths appear.

THE ANCESTORS OF HUMANKIND & THE DAWN OF CIVILIZATION

5 to 7 million years ago: African apes diverge into two lineages, one was gorillas and chimps; the other early human ancestors. (The genetic composition of the great apes is over 96% equal to that of Homo sapiens.)

3.2 million years ago: Lucy, the upright-walking, small brained hominid.

1.8 to 2 million years ago: the emergence of the genus Homo habilis.

1.7 million years ago: emergence of Homo erectus and tool making.

350,000 years ago: emergence of Neanderthals.

200,000 years ago: the genus Homo appears to look like modern man.

160,000 years ago: early Homo sapiens appear.

50,000 years ago: fully functioning Homo sapiens ("wise man") emerge.

12,000 years ago: the beginning of civilizations.

5,500 years ago: invention of the wheel.

5,000 years ago: development of writing and mathematics.

To the present: consult your history books.

What is truly remarkable about evolution is the enormous amount of time that has occurred from the beginning of our universe to the formation of intelligent life on our planet. How can we grasp the duration of one million years or one billion years? And during the billions of years it has taken for humans to emerge, try to visualize the events and forms that coalesced to make things happen the way they have and will continue to be.

We cannot begin to imagine such a concatenation of interdependent phenomena that created such a startling and wondrous universe and a planet upon which millions of different species exist. It truly blows the mind!

The quantum universe

When we deal with the smallest forms—the atomic and sub-atomic particles—we witness a duality of nature, where particles can be particles or particles can be waves, a truly bizarre phenomenon of matter and energy. We also discover that matter builds upon itself in a specific (or quantized) manner so that there exists no continuous or arbitrary multiple but rather a discrete multiple of a physical constant of an exceedingly small size.

The quantum world is for the most part counterintuitive. There exists an uncertainty when trying to measure a particle's location and velocity, whereby when we measure one of these properties, we are faced with probabilities of measuring the other, without any precise results. The observer and his tools have a key role in the result of any investigation of the quantum world.

We also find hundreds of particles that momentarily exist before changing into other particles or disappearing. We also predict that each of these particles has an anti-particle.

The quantum theory suggests that forms and processes exhibit uncertainty and can be found to waver between one way and another, and that in some instances an action in one part of the universe can instantaneously affect another action in the universe no matter what distance is involved! This certainly suggests the presence of interconnectivity throughout the universe. This means that nearby events do not only have local effects but universal ones as well; and, conversely, events of the solar system, the galaxy and the universe can affect all things, including our planet, its geosphere, atmosphere and biosphere.

Because of the theories of quantum mechanics, the ancient expression, "As above, so below," might actually have some validity!

What is Relativity Theory?

The whole nature of physical existence changes when we consider the fact that space and time are related to one another as a curved spacetime entity, where the more massive an object is, the greater its gravity and thus the more it bends light. The enigmatic dark matter also acts like gravity and bends light. In this theory, spacetime represents the entire universe.

A key concept of relativity is that physical laws are the same everywhere and for all observers. Regardless of where a person is, or how fast a person may be moving, the basic physical laws are unchanging. (Perhaps if there are other universes, this may not be the case.)

Special relativity shows that mass and energy are equivalent, which means that a small amount of matter can unleash a tremendous amount of energy. It also shows that energy can create matter, as in the beginning of our universe. Simultaneous events will not appear the same to two observers who are moving in different directions. Time can move more slowly or quickly depending upon whether one remains stationery or in motion.

Also, nothing can move faster than the speed of light. (The exception here is space itself which has no speed limit; thus the inflation that occurred promptly after the Big Bang allowed a stupendous expansion of space much faster than the speed of light. In today's universe, the apparent rapidly expanding galaxies result from the expansion of space, much like the distance between dots on a balloon rapidly growing in volume. A recent study in 2011 has suggested that neutrinos may indeed travel slightly faster than the speed of light. If true, this finding would certainly create a puzzling challenge to theoretical physicists!)

General relativity covers gravity and its effect upon light; timespace; the expansion of the universe; precessions of planetary orbits; and time dilation, including the theoretical notion that if we were to travel in space at speeds close to light speed, we could time travel, leaving the Earth in the 21st century and returning centuries latter without having aged much at all. (Of course, we cannot travel near light speed as our mass would expand with the increasing velocity and suffer dire consequences!)

In terms of everyday activities on Earth as well as the solar system, other than nuclear reactors and particle accelerators, the physical and mathematical laws developed by classical and celestial mechanics serve as reliable guides to deterministic forms and processes.

Manifestations of existence

Actions, behaviors, structures, forms, events, energy—all appear to be manifestations of determinism, chance, synergy or unconscious or unknowable phenomena. We cannot really be certain that what we take as face value is actually what it appears to be. An adequate wisdom acknowledges that when it comes to understanding life and all of existence, we cannot be certain of many of our beliefs. But despite this uncertainty, we utilize common sense and seem to be able to create adequate ideas by which to live and understand things.

What about the physical constants and mathematical certainties? We know that visible light is a manifestation of a specific wavelength along the electromagnetic spectrum. We know that musical notes are a manifestation of certain cycles per second of sound waves. We know that sunrise and sunset are manifestations of the rotation of the Earth. We know that the Sun's rays are the manifestation of the fusion of forms of hydrogen that release astounding energy. We know that the seasons result from the revolution of the Earth and from the angle its axis forms along the path around the Sun (the ecliptic).

Many everyday events and longer-trend activities are manifestations of a wide variety of recurring forms and processes impacting upon our lives and that of our planet.

Existence is the manifestation of all forms and processes interlinking and in concert or at odds with one another. Adequate wisdom is not a piecemeal description of the world; it is rather a holistic extravaganza that requires us to understand in a broad stroke the seemingly endless streams of forms and events that make humans and the universe what they are.

Evolution, design, synergy and God

When looking at the broad sweep of cosmological evolution, we see a moving forward and a trend toward the synergistic assemblage of forms. For example, from the very early beginning of this particular universe, quarks assembled to form protons and neutrons which coupled with electrons to form a vastly inflated universe of hydrogen and helium.

Extremely large clumps of matter, mostly hydrogen molecules, formed gigantic clouds that through the force of gravity and pressure collapsed in such dense states as to create a fusion of hydrogen into helium and other atoms, giving birth to stars and galaxies.

At each stage, prior structures (such as quarks) combine to form another structure, say a proton, which takes on a completely different character. When protons, neutrons and electrons assemble together they take on new forms.

This is the additive nature of synergy.

The critical question raised here is whether or not chance alone was responsible for the assemblage of atoms and stars and galaxies, or whether a blueprint for cosmological forms existed prior to their assemblage—a design. If the latter is correct, then evolution might move forward toward a state of equifinality, which means that there may exist end-states which can be reached by many different means but yet the final form is defined in advance by some built-in archetype.

This design possibility is so maligned by the scientific community that it makes one wonder whether they protest too much about this proposition. There is no proof that physical and biological evolution proceeded only by chance.

A design universe may have no need for a God to create the blueprint. The evolution of forms and processes could emerge from a plan built into the very mathematical laws that direct the determinism creating particles, atoms, stars and galaxies.

As for biological evolution, the vast majority of thinkers believe that chance and determinism alone created life forms. There is another, much smaller group claiming that intelligent design is responsible for life forms (and assumedly all other forms and processes in existence.) In this latter view, the "intelligent" part of intelligent design is meant to claim that God is the prime mover of all things.

It is important to point out that if we accept God as the designer of all things, then we are subjected to an incredible number of human-generated myths, non-sequiturs and phony moral directives, since humans ascribe all sorts of rules and rituals to a God that is totally unknowable. It is the word of a human being who makes proclamations about the revelation he or she experienced in a "conversation" with God. The believer then creates half-truths for practitioners of the faith.

In some cases, Mother Nature might have been the source for religious revelations during the historical periods when and where these took place, mostly in areas of the world where hallucinogenic plants propagated.

Leaving out an entity (such as God) that might have designed the universe, we are still faced with the question whether chance or design is the foundation of evolution. This is where the idea of synergy might have something to say in this matter.

In a chance-based universe, would all things tend to form greater and greater wholes? Is it just chance that formed the 92 natural atoms along with hundreds of isotopes, isomers and ions? Is it merely chance that created the carbon atom allowing it to form a multitudinous number of organic molecules? Is it chance alone that created primitive living cells only to evolve into greater and greater wholes to form tissues and organs and organisms—from gnats and giraffes to human beings? Is it chance alone that created the hierarchies of animal and plants?

(It should be pointed out that chance, of course, plays a vital part in cosmological and biological evolution, allowing for the incalculable diversity of stars, galaxies, planets and life forms.)

Synergistic wholes abound throughout the universe. Do these wholes represent only chance and determinism, or do they represent some form of design? Foreordination is a concept alien to most thinkers. It smacks of a God and therefore obscures the possibility that the initial or later conditions of the universe might have held built-in instructions, not necessarily planning every aspect of evolution but perhaps directing it along with equifinality toward synergistic wholes.

This is of course quite speculative, but cosmic design without an omniscient God is an idea to consider rather than discard.

The pulsating universe

Both structures and forces exhibit motion and activity in the form of vibrations. From the smallest form to the largest, there is constant energy manifested in spin, revolution, precession (wobble), electric and magnetic oscillations. All matter consists of particles and electrons, and these are in constant motion.

The constant wanderers are the electrons as they jump about in a fuzzy manner around the nucleus of all atoms.

Even the so-called vacuum of outer space vibrates with the creation of virtual particles that come and go out of existence on a continual basis.

Nothing in the universe is motionless.

Because of the quantum nature of things, that is, because energy comes at us in discrete packets (photons or radio and television signals) and not continuously, forms give off specific signatures. Why is an object red and another one blue? Because of a specific oscillation along the electromagnetic spectrum. When playing a musical instrument, why is one note a B flat while another is a D sharp? Because of the specific oscillation that each one generates.

Not only do particles, atoms and molecules pulsate, but all things pulsate, including living cells, organs and organisms, each with its own unique signature.

The strange relationship
between form and process

We have said that form is structure, like the atom or the human being. And we have said that process, like evolution and consciousness, is change and novelty.

In physics, mass is a form, like an atom. But within that form, there exists an enormous amount of potential energy, so there is some equivalence between form and process. Gravity is a force, a process, yet gravity (along with dark matter) also serves as a structural agent, keeping stars and galaxies together. In theory, gravity as a force is represented by massless particles called gravitons.

Electromagnetism is a process, yet it has structure based upon its range of wavelengths, which can be construed as specific vibrating forms.

Matter exhibits both particle (form) and wave (process) duality.

Can form and process be different aspects of the same principle? Previously in these essays, we have mentioned that form (structure) can direct process (function), while process can direct form.

Form and process seem to be inseparable as are all the dynamics of existence.

Prediction and uncertainty

Some events in this world are predictable, which means we know precisely what will occur before they occur because of prior observations. Take for example the combinations of elements in the Periodic Table. We know that a precise mixture of sodium and chorine will produce common salt. The mathematical and chemical laws are very precise for many of the 92 natural elements and their various combinations and bondings.

The primary physical and mathematical constants of the world are precisely measureable, although some exhibit a very slight variance.

Also precise are the mathematical equations of celestial mechanics, so that the motions of the Milky Way, Sun, Earth, and other celestial bodies can be predicted precisely for hundreds, if not thousands of years ahead (or in the past). We actually can tell the future of celestial motions with amazing precision. We know when high and low tides will occur, when eclipses will occur; and we know where each and every planet is located at any given moment in time—past, present and future.

Prediction is a hallmark of determinism—of science, technology and mathematics.

When we look at the particles, predictions are based upon probabilities. High probabilities exist, but are less than 100%, thus creating uncertainty.

In the world of genetics, we know that certain genes call for the creation of certain proteins; we know what chromosomes carry specific genes. But we cannot account for mutations, mistakes and recombination, and thus there is uncertainty in the creation of a genotype. However, there is a high probability that certain genetic flaws will cause certain dysfunctions in the body. Modern medicine now has the ability to ascertain some future defects of offspring by studying the blood and tissue of the mother and fetus.

In biology, we can predict with accuracy the functions of organs and organ systems, yet we are uncertain of the individual's fate in the world.

Uncertainty is prominent in many events in the mundane world. Unpredictability finds itself lurking in everyday human interactions and throughout the lifetime of each and every human being. Diseases associated with undiagnosable defects in the genes and the organs are unpredictable. The stock markets of the world, while certainly affected by good and bad economic conditions, still offer uncertainty, as most investors will attest. Every sporting event is an example of uncertainty since we can never predict with accuracy the actions and results of the game. An incalculable number of factors interplay with one another and create unpredictable results in every manner of human social intercourse.

Human beings, individually or in groups and institutions, do not like uncertainty. Therefore, most are adverse to change and resist it. Likewise, most humans do not like surprises, unless of course they present good news.

Many political and social conditions in all nations are unpredictable, as they are based, in part, upon the oftentimes irrational and emotional nature of humans, who themselves are subject to chance as well as free will. The unpredictability of events and relationships among people and institutions make life interesting, sometimes very pleasant, other times frustrating.

But fixed laws, customs and regulations exist to promote stability and certainty. Throughout history, there has been a push-pull effect, whereby forces exist to promote a civilization in equilibrium vis-à-vis the forces that pull it apart.

Predictable events are based upon determinism. Uncertain, anomalous and/or irrational events are based upon chance, free will and the unconscious mind. Thus in the course of everyday events, there is little certainty that can be assigned to human interactions.

Will there be an end to the current universe?

Conventional wisdom about the end of our perceived universe depends upon these possible scenarios: (1) the universe will continue to expand and then reach a point where it dies as all energy has been expended; (2) the universe will begin to contract and fall back upon itself until it returns to its initial state; or (3) the universe will continue to exist by teeter-tottering between expansion and contraction.

Right now, the universe is in a major expansion stage, because the so-called dark energy creates a repulsive force at breakneck speeds, so that the universe may indeed approach a final heat death. But given the eons to follow, nothing says the universe might not return to a contracting stage, and so one must be careful in assuming a non-stop expansion in the far future. So in a trillion years, our universe might coalesce into a giant membrane which would collide into another membrane and create another universe, as suggested by string theory.

If we believe that only one universe exists (ours) and it may possibly end—well, that's a bit too final an option for most of us to accept. How can we grasp the notion that our universe was born from nothing and ends in nothing? How can the human mind even begin to understand *nothingness*? Which is why we tend to believe that even if our universe were to end, there may be an untold number of universes existing ad infinitum, thereby sustaining the idea that there is neither beginning nor end to existence.

The implications of infinity

If there are 10 space dimensions according to string theory, seven of which are compacted to form an infinite number of universes, and one time dimension, then everything we have believed about a single universe is wrong. It is estimated that there could be 10^500 or more universes formed by strings and other structures postulated by M-theory. This number is so large as to constitute a virtually infinite number of universes.

What this suggests is that everything is possible. There could be an endless number of universes that operate under an unlimited number of physical forces and matter combinations. Taking this a bit further, since everything is possible in an infinite world, it is very likely that you and I and all humanity, stars, galaxies and elements might be replicated in another universe by some of these unlimited possibilities –that every action we take is replicated elsewhere as well as an endless amount of variations of ourselves and our actions and lifetimes! So-called parallel universes may exist and present us with this unbelievably staggering idea. Yet this is possible in an infinite world.

With an infinite number of universes, chance and determinism establish their own particular set of physical laws which then become determinants of forms and processes. Within some universes, however, there may be design built-in, so that certain structures emerge, such as atoms, stars and life forms; and they in turn form synergistic wholes. Perhaps some universes contain no life forms at all. Perhaps some feature living things with intelligence but with unusual personality and behavioral traits. In other universes, there may exist phenomena about which we could not even venture a guess.

There is also the possibility of universes within one universe, so that what we hypothesize as multiple and infinite universes might actually exist as one eternal existence, with big bangs as merely hiccups in the grand synergy of forms and processes within a single, everlasting whole which we might call eternity.

Adequate wisdom and physical existence

Assuming that form, process and idea, as well as design, determinism, chance, synergy and free will constitute the key factors in the quest for adequate wisdom, then what conclusions can we reach about existence?

We know that forms are structures or wholes that permeate the universe. We know that processes are the forces that interact among and within wholes. We know that wholes subsume smaller wholes and become larger, unique wholes. We know that free will permits us to create ideas about the forms and processes, while volition permits us to make decisions, take actions and create things and policies.

We know that some things are designed and determined, while some things happen by chance. So the interplay between form and process establishes a basis upon which to build an idea system to guide us in our search for wisdom.

What forms are determined?

We know that gravity helps to create galaxies, solar systems and stars. We believe that strange dark matter also helps to hold the galaxies together. We know that the strong force holds protons and neutrons together to sustain the nucleus of an atom. And we know that the electromagnetic force guides the electron along its fuzzy orbits and sustains the 92 natural chemical elements, many of which can be manipulated and/or manufactured into a plethora of molecules and polymers and many new chemical compounds.

These forces are determinants—they sustain the forms and mediate among them. An atom is a determined structure which exhibits such processes as spin, charge, revolution and others. So while an atom is a form, it is also defined by the processes.

But, we might have to revise our thinking if the "chicken" precedes the "egg." Perhaps it is the form that determines the processes. The blueprint for a galaxy or an atom might have existed prior to the existence of gravity and electro-magnetism. This would mean that the key forces in the universe are preprogrammed. The concept of preformation cannot be denied as one possible interpretation of exis-tence. And, even more confusing, perhaps form and pro-cess are equivalent, much like matter and energy, so that evolution and preformation are equivalent as well. (Talk about counter-intuition!)

We are then forced to admit that our thinking remains fuzzy about existence. Most people believe that cosmologi-cal evolution occurred in a linear manner, from cause to effect. This is backed up by the direction of time or evolu-tion, which seems to flow ahead and perhaps finally reach a heat death. Linear evolution seems to be the call, but we must not rush to a final judgment on the matter. If we 100% guaranteed that evolution is linear, then we would rule out every other possibility without truly understand-ing the universe.

We conclude that any designed whole is highly determin-istic. We know that it comes into existence based upon preprogrammed schema. The erection of a building follows strict instructions from its blueprints, so the final end has been based upon deterministic design. Chance plays its part, since some steps in erecting the building may not have been followed exactly and therefore cause flaws or weaknesses. The terrible tragedy of the complete collapse of the World Trade Towers was in part due to inadequate bolts—of course, the designers could not have ever con-ceived of a terrorist attack, and thus chance events con-tributed to the disaster.

What if preprogramming does not exist? Then how do par-ticles and waves come into existence? How does gravity come into existence? How does the strong bonding force exist that holds together the nucleus of an atom? How does the weak force allow for the decay of elementary particles and for radioactivity?

Physicists can ascribe mathematical formulae to these forces and shrewdly show how they work together to form atoms, photons, stars, galaxies and the chemical elements. And all these physical forces can be measured and shown to be based upon determinism (and possibly by the theoretical vibrating strings).

It is best to assume that forms and processes are interdependent. They can be deterministic and chance-based, as well. Any mundane event is the result of form and process and the interdependence of free will, design, determinism, chance and synergy.

The trend towards synergy

The combination of simple carbon-based compounds into living forms is one of the greatest synergistic events in existence. And the synergy continues on and on, through the combination of amino acids (proteins), organelles, nucleic acids and complex cellular structures that created millions of living species and newer species to follow in the near-future or in hundreds of thousands or millions of years.

The hierarchy of biological and physical forms, from simple cells, tissues, organ, organ systems, organisms, kingdoms, populations, social institutions, the biosphere, the hydrosphere and the atmosphere—all exhibit a synergy which signifies that we humans cannot operate under the illusion that we stand alone or that we are the only significant forms on this planet. Each of us is part of our specific species and each human is subsumed by the biological destiny, chance conditions and free will that make us what we are.

In the social sphere, synergies abound, as individuals form small groups (mates, friends, lovers), larger groups (clubs and organizations), institutions (schools and universities, corporations, military units, religions) and so on—all individuals subsumed by higher order structures and somewhat molded by each particular whole. Each new whole carries out its own particular processes.

Many nations form larger associations, like the United Nations, SEATO, OAS, NATO, the World Court, the World Trade Organization, the World Bank and many more collectivities. If the trend towards synergy stays true, then it is possible in the future for the entire human population to become part of the world community. If this does occur, we are speaking about a very long time period, since, like individuals, some nations guard against collectivities, preferring to go it on their own without "outside" regulations.

In the physical world, two varieties of quarks combine in different relationships to form protons and neutrons which become new synergistic wholes. The electrons and their expanding relationships with protons, along with neutrons, constitute the 92 stable atoms or elements in the Periodic Table. When neutrons are added to the stable elements, there comes an assortment of new synergistic wholes (isotopes). Most of these atoms form an incalculable number of relationships with one another to form all kinds of synergistic wholes (molecules)—the stuff of the world, so to speak.

Atoms stripped of their electrons become ions and form plasma that occurred in the beginning of our universe. When temperatures cooled, the particles combined to form hydrogen and helium which eventually formed clumps of enormous molecular clouds that gave birth to stars and galaxies and light. *It was the synergy among quantum fluctuations, gravity, electromagnetism, strong and weak forces, dark matter and molecular clouds that formed the stars and galaxies.*

Although the specifics of galaxy formation are not understood, it seems possible that the collection of stars of various sizes and magnitudes combined in various arrangements to form the synergistic whole—the galaxy. Galaxies grew in size as they cannibalized or combined with other smaller galaxies, thus becoming a new form. Galaxies also gathered together to form groups, clusters, superclusters and super-gigantic walls or sheets of clusters.

It is interesting to note that even molecular clouds that give birth to stars are arranged in a hierarchy because of gravity and shock waves which create a set of sub-structures in the cloud in hierarchical order.

Toward the end of a star making process, as the new star spins, it spews forth gasses and heavier elements which coalesce to form bodies circling the new star, including all sorts of rocky or gaseous structures that eventually form planets, moons, asteroids and other astronomical objects.

The synergistic trend began with the combinations of particles and atoms and continues through the present time with the formation of new stars and new life forms, including human beings. Based upon some ideas of quantum mechanics, it is quite possible that the universe and all of existence is one Grand Synergy—a collection of parts and wholes mostly operating beyond our comprehension.

ADEQUATE WISDOM

————

PART SIX:

————

GOD & RELIGION

Part Six provides a wide range of essays about the notion of God, atheism, agnosticism, the soul, prayer and the phenomenon of religion and its powerful force in the human world; including its influences upon human sexuality.

What is meant by God?

As a concept and a semantic tool, God does not have to be associated with any religious belief. God can signify all that is unknown. This semantic God has no resemblance to anything human and is not concerned with the plight of humanity. It may simply represent the unknowable or inexplicable essence of existence.

There are those who look to the physical universe as God, as well as marveling at the very existence of nature, yet denying the existence of an all-knowing, humankind-directed entity. God may signify the preformational nature of existence, meaning that, like an acorn, the universe unfolds according to a preprogrammed plan or design. In other words, the entire "gross" features of existence may be pre-ordered, whether by mathematical and physical laws, or by something well beyond our understanding, which can be called *God*.

God can also signify one or more divinities that are believed to create and/or regulate the world, including human events. This is the hands-on, human-generated God or gods in whom a vast majority of humans believe, now and through all history. It may be one God (as in Christianity, Judaism or Islam) or a variety of gods (like the many Hindu or ancient Graeco-Roman gods).

Some philosophers and theologians claim that God created everything, but allowed humans to have free will. This two-tiered approach to God seems like a convenient way of explaining existence, but can we really accept the idea that a divinity created and managed the universe, the galaxies, stars, planets, life forms and all the physical forces, and then suddenly stopped managing in order to offer free will to humans?

It should come as no surprise how manipulative this concept is. It conveniently explains why good and bad things happen, since a divinity allowing humans to experience free will and design washes its hands with human behavior and places the plight of humanity in its own lap. (The irony here is that even though many religions espouse free will, religious dogma violates the free will concept by telling us what God expects of us and prescribing manners by which we live—all in direct opposition to the concept of free will.)

What puzzles the rational mind is the fact that a vast majority of the population believes in a God that watches over us, and because this God supposedly answer prayers, followers practice a wide variety of rituals that are based solely on human design. There is no meta-entity calling out to certain people and giving them laws and edicts which they must follow. Humans have invented the idea of a divinity, as well as the curious, enigmatic, benign and, at times, harmful practices of religions.

However, to create the concept of God as an eternal entity (form) representing infinity seems to be a legitimate idea when juxtaposed with the concept of evolution (process), allowing us to offer a partial reconciliation between religion and science.

Using God as a representation of the entirety of existence—all the components working together with and without our understanding—might include such phrases as:

"God bless you." This could simply mean that you hope many of the forms and processes in the world will work mostly in your favor. It can also be used to thank someone for doing something good for others.

"God help us!" This is simply an expression of hope that forms and processes will provide relief from stressful or terrible conditions.

"God forbid!" This expression simply means that we hope that something terrible does not occur in the future.

"Thank God!" Here we use this expression as a gesture of appreciation for something good occurring—a positive confluence of forms and processes and ideas.

"God damn it!" Here we express disapproval or disgust with certain events and strike out at the swirling forces around us that have conspired to cause frustration, grief, pain or stress.

In these examples, the use of the word *God* provides a convenient way to summarize all the forces and forms intermingling with one another. One does not have to believe in a specific divinity to invoke the word *God*.

Does God Exist?

One of the most intriguing questions ever raised, whether God exists has been at the forefront of religious and philosophic thought from the beginning of humankind. Humans look out at the heavens and see a most spectacular universe or look within to see the extraordinary workings of a tiny cell that unfolds into a living creature that is capable of asking questions.

Many people assume that existence has a designer, something that created all that is. Belief in a God attempts to offer a clarification of the confusion that most mortals hold about the vicissitudes in their lifetimes.

But can we really know for certain that God exists?

The most certain among us claim that God exists based upon revelation and faith. This God has advocates who claim that a divinity has a very specific influence in the lives of people. Paradoxically, others claim that God exists but grants free will to human beings.

Others are not so sure God exists, and will understandably equivocate and say it may be possible. Some of these people may believe in some unknowable entity or force, but do not necessarily believe in a God that directs humanity.

A few others do not believe in God at all. They believe that chance and determinism and other factors are responsible for all that exists.

The possibility of an infinite world, where spacetime is endless, might preclude the need for a designer or God, for there would be neither beginning nor end. The idea of eternity, which has been bad-mouthed throughout the modern scientific age, has gained support from theoretical physics whose M-theory and quantum theory allow for an infinite, eternal universe.

Whether a human being believes in God or not—his or her life plan can be fulfilling and enjoyable if one is open-minded and compassionate.

However, the danger of believing in God comes from those who pontificate all kinds of laws and edicts that are ascribed to such a God. How many times do we hear the phrase, "The Word of God?" The word or words of God are attributed to religious texts or revelations written or spoken by humans who truly believe they speak for an unknowable God. Frankly, the concept of God has been abused from the beginning of human thought. It is used to coerce and dominate people. It is used to instill fear and reverence for something that no one can ever comprehend.

It seems quite understandable for people to sustain a faith in something bigger than life, bigger than the cosmos; but it appears counter-productive to follow the rules, regulations and moral codes of an ineffable God, whose creeds are solely based upon the ideas of human beings.

Belief in God is an admission that most members of our population place their trust in an unknowable entity which somehow looks after them. Religious people place themselves in a subservient role, and rather than follow their own common sense and wisdom, they often place themselves at the mercy of an unknowable entity as well some questionable religious edicts.

Belief in God is an acknowledgement that most people need to have faith in something which allows for a purpose to life and an everlasting place in eternity.

And this belief is all well and good if God is not vengeful nor vindictive, but compassionate and kind, promoting the success of all living creatures and not beholden to any particular morality.

Atheism and Agnosticism

This work does not promote the belief in atheism, which by its very nature denies the possibility of some unnamable factor or force which may guide the universe. The synergistic and holistic nature of existence allows for the possibility of forms or forces beyond our perception. Therefore, to deny categorically that something exists beyond our comprehension is the height of conceit.

But common sense dictates that no one can speak for a God and no individual or institution can tell people how to behave and decide what is right or wrong or evil based upon such an unknowable entity. Nonetheless, belief in something greater than the individual parts of existence should not be ridiculed.

Individuals who profess to be atheists claim that there exists no such thing as a soul or no such entity as God, and of course this is their opinion and they should never be admonished by those who believe otherwise.

A sensible viewpoint, according to adequate wisdom, is agnosticism, which hedges the bet and allows for the possibility of a God or some unknowable force throughout the universe. A true agnostic would confess to his or her incomplete knowledge of existence and suggest that the atheist and the theist cannot support their views with credibility.

The possibility of God representing the mysteries of existence is acceptable in adequate wisdom, but all the embellishments and dogma created by a belief in God hold no standing among rational people.

Adequate wisdom based solely upon a divinity

Those who subscribe to the belief of an omniscient God explain existence as the result of a divinity creating all that was, is or will be. This concept is sometimes referred to as creationism. Most versions of this belief leave no room for evolution, chance and free will. Creationists have a difficult time explaining why God took about 10 billion years to create the first life forms on Earth.

The thesis that God accounts for everything is a convenient way of saying that everything is God, a tautology if ever there was one! But if we strip away all the religious dogma and declare that God may represent eternity and all those puzzles we cannot comprehend, then we can allow the use of this terminology because we probably do need some name to reflect our bewilderment as to the complete nature of existence.

Having said that, however, we cannot reach any truth because this God is unknowable, ineffable, and any claims made on its behalf are merely invented—the result of humankind creating innumerable religious edicts, rules, prayers, regulations, commandments and admonishments in its name. Therefore, a series of questionable moralities are created, telling people, in God's name, how to act, what to think, what to wear and whether, in some instances, they will go to heaven or hell.

Many religions want their cake and eat it too, so they claim that God has given humankind free will to do what it wants to do, but always under the umbrella of a morality designed by humans but beholden to God's will, whatever that may be. Thus, we do see that systems of morality can be devised, and that life can be given meaning, as it is defined by each religious system. However, the religious belief systems and their creeds and dogma cannot rise to the level of adequate wisdom because they do not incorporate the variables of existence and do not acknowledge the dynamic relationships among forms, processes and ideas.

Does the soul exist?

Many religions consider the soul as the foundation of each human being. It is thought to exist outside of the material body and mind, and therefore an essence unto itself. The soul is supposed to transcend form and process and therefore cannot be destroyed upon the death of an individual. In some cases, the soul and the spirit may represent the same concept. There also exist fine distinctions between the two.

Without the concept of soul or spirit, religions would be hard pressed to explain death and an afterlife. The absurd notions of heaven and hell are merely a strong-arm threat to intimidate people into following arbitrary moral directives.

We must also ask why other living things are often thought to exist without any soul. What would make humans worthy of a soul and other living forms unworthy?

Belief in a soul requires the elimination of two of the three tools of adequate wisdom—that is, form and process, leaving us only with the idea of a soul.

Can we extend the definition of soul to suggest that everything in existence is represented by an essence beyond our comprehension? The concepts of animism and vitalism generally suggest that all forms have souls or essences beyond the grasp of the rational mind. Are there unknown fields that are responsible for the creation of matter and force?

Since adequate wisdom is based primarily on rational and commonsensical principles, the concept of soul does not appear credible, especially since credible ideas are generated by the interaction of forms and processes, and the soul exhibits neither of these.

The soul's supposed immateriality, however, places it squarely into the metaphysical arena in which intellectual battles are often fought without anointing a victor.

Why do religions exist?

The most important reason for religion to exist and flourish for the past several thousand years is that it offers explanations to those who exhibit fear of the unknown, to those who want answers about the key questions: why do we exist; is there any meaning to life; how should we behave; and what, if anything, happens to us after we die?

Religion has institutionalized the belief in a supreme being and has established a variety of labyrinthine bureaucracies that combine myth, lore, revelations and ritual—all based upon ideas created by humankind and not from divinities.

For most people, religion provides a way of life, of beliefs, activities and social interactions. When religious groups care for the downtrodden, they are providing a compassionate service to humankind. When they construct and perpetuate moral systems that instruct people how to think, act, dress and behave, they perform a major disservice to humankind.

Moral guidance for our species should come mainly from rational, secular laws and common sense traditions, and not exclusively from religious edicts. If we follow only the moral guidance of religions, then we are faced with a variety of contrasting and disparate indoctrinations, many based upon fantasy and misinformation.

Earliest man—before recorded thought—was certainly frightened by and in awe of the spectacle of storms, lightening, earthquakes, volcanic eruptions, eclipses, sickness and death. Early man thus created gods to help explain phenomena and provide some means of succor. Indian religions (eventually building into Hinduism and later Buddhism) are thought to be the first on the planet, followed by Judaism. The pre-classical period in Greece and Rome featured many gods and rituals, but cannot actually be classified as religions.

The advent of Christianity and then 650 years later, the rise of Islam, round out the major religions in the world today.

Religions exist because of the frailty of the human condition—sickness, loss of a loved one, terrible diseases—as well as the wonderment that most people express when observing the marvels of life and the cosmos—the so-called WOW! factor.

Another obvious reason for the preponderance of religion is the fact that it brings people together and occupies their time during their lifetimes. For a large majority of people throughout the globe, religion is a way of life, a social venue, without which their lives might feel empty and without purpose.

The success of religion over any other alternative form of belief (such as science, common sense or humanistic philosophy) is based upon its oftentimes simple explanations that comfort people. When a person dies, religion offers calming and reassuring guidance to the grieving family that the soul of the departed will remain for eternity. Contrast this approach to a non-religious one that explains death as a cessation of all living matter with no soul. The latter may be correct, but the former offers much greater comfort.

Religion has been rigidly institutionalized so that it is rarely given to changes and reforms. Many established and dominant religions have created an unbendable set of rules and regulations.

The greatest strength of religion is that it offers solace to its members in times of disaster and grief. Its greatest weakness is its inflexibility to change. It locks itself into moral rigidities which at times create a despotic regulation of its members, leading to zealots who have upon too many occasions rendered despicable acts against humanity.

However, there are many good people who serve in the religious orders and who provide kindness and support for individuals whose lives are racked with discomfort and pain.

Religion has inspired the creation of great art, music and architecture; has contributed to literature with endless allusions to religious texts; and provided a way of life for billions of people throughout the ages. It is an extremely powerful force in civilization. There are indeed positive messages from religions, such as helping the poor and the unfortunate; but, alas, these messages are often directed to members of a specific religion and are not always worldwide in their application.

If secular beliefs can offer rational explanations of existence while providing comfort to those in need, then a new religion devoid of alleged miracles, falsehoods and myths might arise over the next millennium.

How does religion affect a life plan?

Religions offer a wide variety of creeds, rituals and systems of rewards and punishment, many of which are inflexible. One who seeks to develop a life plan would be best served by examining the beliefs of religions and selecting those aspects of religion which seem commonsensical and incorporate them into one's own life plan.

A nearly universal belief of the major religions is to care for those who are less fortunate than we are. The idea here presumes that our species is a collectivity of all members, and thus caring for others makes good sense. The concept of compassion is a core belief of most religious institutions. Unfortunately, it must be stated that compassion for competing religious groups or non-believers is not always promoted.

Religions promulgate different moral systems, so one must be careful to examine them carefully, since these systems often demand exact behaviors. The life plan must carefully weed out those edicts that undermine a commonsensical approach to life. Blind faith is extremely dangerous—it encourages arbitrary rules and regulations, many of which actually do harm to others.

As we all know, many of the injustices and wars inflicted upon humanity have been triggered by religious zealots, religious fundamentalists and by fanatical moralists whose ideological inflexibility has been the bane of existence.

As for the idea of an ideal religion or system of belief to be incorporated into a life plan, one would expect rationality, compassion and the principles of adequate wisdom to be observed.

What is Prayer?

To whom or what do people pray? It is to a god or the God or divinity chosen by any particular religious institution. Prayer may be performed by an individual person, small groups of people or large assemblages of persons. In a "low church" approach to religion, worship usually occurs between an individual and his presumed maker without pomp and circumstance. In a "high church" approach, all the bells and whistles of the religion are added to the prayer ceremonies.

Whether or not an actual deity exists, prayer serves as a useful psychological tool for those engaged in it. It is very common for people to pray for the recovery of a victim of a terrible accident or physical disability. It is common for members to pray for the health and well-being of their families, by imploring a divinity to grant special benefits to them. Prayer brings people together and helps to solidify a social group.

Public prayer is based upon highly structured and ritualized proceedings with rich pageantry, often using texts that are as old as the religion itself, some of which having been modified to suit the current age.

Prayer is the manifestation of people's admiration and awe of a higher entity. It signifies that each member of a religion believes that he or she is not totally in charge of one's actions. It is the acknowledgement of an unnamable, ineffable power or force that may guide our lives.

In a disturbing trend, some nations have prohibited the wearing of Muslim head coverings worn by women in public. The city of Paris has prohibited public prayer by Muslims. Whether or not one agrees with religious customs, the selective suppression of a particular religion's rites works against an open-ended, free society.

What can be said about those who do not pray? Obviously, they do not believe in any particular prescribed religion or religious text. It does not necessarily exclude them from believing in some mysterious or inexplicable universal unity. Their lack of prayer does not exclude them from exclaiming "Oh My God" when special events occur in their lifetimes. The positive "Oh My God" moments occur when amazing and surprisingly good events happen. The negative "Oh My God" moments occur when terrible disasters or tragedies happen. (These good and bad events align with the confluence of forms and processes.)

The very cynical or detached members of our population believe that all events, no matter how spectacular and amazing, are attributed to mere physical activities in the universe. They scoff at any notion of a universal force that may direct the evolution of atoms, stars or living beings. They also believe that prayer is useless and non-productive.

Applying the principles of adequate wisdom, with its observations of forms and processes, and the modes of design, chance, free will, synergy and determinism, we may conclude that there is the possibility of a goal directed universe. Because of the synergistic trend of physical and biological evolution, we cannot rule out with certainty the existence of something well beyond our comprehension.

On the other hand, praying to an incomprehensible divinity or force does not appear to bring any results, except for psychological relief.

Therefore each human and every social institution must create wise policy decisions themselves, without depending upon a personal or transcendent god to help.

How do religions treat human sexuality?

The monotheistic religions have an odd sense about pleasure, especially human sexuality. In essence, many of them tend to disapprove of the sexual pleasures of the body, except for use in propagation of the species, which of course is essential for perpetuation of the species. But when it comes to non-marital sexuality, they often believe it has harmful effects on humankind.

The seeking of non-marital orgasm and the innumerable paths to reach it are regarded as questionable, to say the least. Sexuality has always been a near-taboo subject in Western religion, a subject which engenders a form of shyness or disapproval. Put plainly, many religious people are very uncomfortable with the subject of sex. Even secular parents seem uncomfortable discussing sex with their children.

Using questionable passages from their scriptures, religions as a whole regard non-marital human sexuality as a distasteful preoccupation from which one ought to refrain.

The Eastern religions are not as stringent regarding sexuality, especially some forms of Hinduism, but these religions, along with the Western ones, all feature branches of mysticism that primarily tout abstinence from all physical and worldly pleasures—a form of asceticism that dates back to ancient times.

Now it is true that without any self-control, humans may very well act as do animals in regard to their powerful sexual drives, so it is understandable that society creates laws and mores to prohibit improper or extreme acts of sexuality, especially rape and other forms of sexual violence and predatory behavior.

The male sex drive, acknowledged as more aggressive and stronger than the female drive, does require some form of moderation. But the pendulum had swung far to the side of dominant suppression of sexuality in the religious dogma during the past 2000 years. Recently, the tide has turned somewhat. Open societies today allow far more sexual freedoms, overriding religious restrictions.

How do religions and society treat homosexuality?

The fascinating incidence of homosexuality and society's regard for it has also seen various pendulum moves. In the era before the advent of Christianity, it appears as if homosexuality was neither taboo nor one with which the pre-classical era had a problem. The exception is the Old Testament with explicit disapproval and condemnation of acts between the same sex, along with condemnations of many other acts, most of which in today's world seem petty, trivial and harmless.

The Western religious dogma still prohibits and/or frowns upon same sex acts, yet a growing number of religious groups seem to accept it and call for compassion for all sexually active members of society. The Presbyterian Church (USA) in 2011 voted to change its constitution, allowing openly gay people in same-sex relationships to be ordained as ministers, elders and deacons. They join a number of other Protestant church groups (as well as a few other religious sects) accepting and embracing the gay lifestyle.

Hinduism has a mixed stance on homosexuality. Same sex acts are prohibited in the higher caste system for those experiencing the "twice born" rituals, and the nation of India does not fully approve of homosexuality; yet same sex acts are encouraged in the Kama Sutra.

Buddhism has little to say about homosexuality. Where Buddhism is practiced, same sex acts are generally regarded as something like misdemeanors, and there is nowhere the distaste for it as in Western religions, although in certain regions it is still viewed unfavorably.

In many Islamic nations, homosexual acts are strictly forbidden, scorned upon and subject to severe punishment, including, in some instances, stoning, dismemberment or death.

One may venture a guess as to why same sex acts are repugnant to Western religions (in addition to religious texts forbidding them) and that would be that these acts do not produce offspring and therefore, according to these religions, fall outside of the "natural" world. In addition, religions require offspring to sustain and add to the number of a particular religion's population.

Fundamentalist and some mainstream religions exhibit disdain for the celebration of physical pleasure (whether it is heterosexual or homosexual). Chance and determinism also create the phenomena of effeminate men and masculine women, as well as people whose genetic make-up creates gender confusion. These traits are the subject of ridicule and disapproval; yet as compassion grows in our society, these atypical sexual traits are becoming viewed as part of nature and not to be derided.

Is homosexuality part of the natural order of things? Evidently the animal world is filled with examples of species where homosexuality and bisexuality are evidenced. (According to members of the Natural History Museum, University of Oslo, Norway, more than 1,500 animal species have been identified as practicing bisexuality or homosexuality.)

Its opponents regard homosexuality as a choice. Even the slightest contact with common sense should indicate that sexual orientation is not part of free will and human choice but attributable to chance and determinism.

At first glance, it seems that natural selection and genetics are not factors in the phenomenon of homosexuality. Since it does not increase the population, homosexuality should have been eliminated a long time ago. However, as a means to prevent overpopulation, natural selection must actually favor homosexuality.

One of the greatest ironies in the history of American homosexual rights is the disapproval of gay people by a majority of African-Americans, themselves an historically repressed minority, who have nonetheless contributed to the repression of another minority. Much of this disapproval is based upon the inflexibility of the fundamentalist churches and their moral crusade against gay people.

There are certainly other individuals or groups not affiliated with religion who oppose homosexuality. They may simply have a visceral objection to same-sex relations or, by way of upbringing, ignorance or hatred, they have a need to denigrate people whose lifestyle departs from their own.

In any event, those who have been persecuted and humiliated throughout the ages because of their sexual identity or for any difference in skin color, age, genetic dysfunctions and the like, deserve a global apology for their reprehensible treatment, as well as policies and laws to correct the denigration they have received and to provide protections from further abuse and ridicule.

The trend in Western democracies toward homosexuality has been mostly positive; but those in less developed nations still exhibit much prejudice.

ADEQUATE WISDOM

PART SEVEN:

HUMANITY

Part Seven contains a sweeping look at various topics that represent and reflect the human condition and day-to-day transactions among people. Such topics include: egoism and altruism; intelligence and wisdom; compassion and punishment; love and attraction; the influence of sexuality; policy making; absolute and relative values; death; and whether there is any meaning to life.

What influences human beings?

In short, the entire universe influences us. All the cosmic events that occurred to allow life forms to exist play a factor in our lives, including the formation of stars, galaxies, the solar system and Earth. The recurring forms and processes affect each one of us and civilization as a whole.

Let us look at the lifetime of a single human being. He or she starts out as an adjunct to the mother, bathing in her biological fluids and forming according to the genetic codes which give rise to a pre-programmed set of tissues that form distinctive organs and organ systems, including the central nervous system, where the firing of neurons give rise to consciousness.

Some physical characteristics, talents, abilities, tendencies and proclivities of the parents are often passed along to the newborn during fertilization which forms the first step in the development of a unique human being.

Mistakes in the genetic codes can create a series of mild to tragic dysfunctions in the newborn or lay dormant only to form diseases in later life, whether in adolescence or adulthood.

The newborn child is influenced by nutrition and the hospital and its staff. Anecdotal observations point to the fact that the personality traits of the newborn emerge very early on and seem to exist independent of the traits exhibited by the parents or siblings.

Personality and emotional traits are an extremely important influence in the lives of each human. One should never underestimate how much a lifetime is affected by the personality and character of each member of our population.

As the child grows, he or she is subject to the political, religious and social views of the parents, and by their economic and social status; as well as peer groups, all having a strong influence. The child and adult are also influenced by the circumstances in which they find themselves. Obviously, the nation, region, state and city will have rules, regulations and laws allowing or disallowing certain behaviors. The education of a human is exceedingly important for the development of his or her mind. The times in which people live have a powerful influence on the person, which would include the economic, political and social events during the lifetime.

The development of a conscience which determines right from wrong as well as other judgments made by each human are the result of a stream of influences from parents, friends and society (including customs, religion and law, as well as each person's thought processes.)

Most people express the belief that there is no better time to exist than the present, since the past was not as technologically advanced. To cite one example, medical advances are presently available to help people live healthier and longer lives. Nonetheless, some people yearn for the past where technology was not dominant, and therefore people were able to connect to one another more socially. In any event, the era in which people live has a tremendous influence upon them.

The social structures in place during a human lifetime bear a strong influence upon the individual, as families, groups and institutions provide the civilizing imprint upon each of us; and no matter what type of personality or character exists, the individual reflects his or her society.

A great motivator of human behavior is the avoidance of conflict. Except for the aggressive, belligerent personality types, most humans seek to avoid confrontations, nasty verbal and physical clashes and a wide variety of interactions among others with whom conflict seems inevitable. Another motivator is fear, either real or imagined or ingrained by outside forces bent on manipulating people to their ends.

One must never discount the extremely powerful influence of material gain, the quest for the accumulation of money, luxury, fancy cars, richly appointed residences, precious metals, technological gadgets and all other "treasures" abounding. Greed greatly affects and drives the lives of many members of our population.

The sexual drive within each human being can be over-whelming and at times overshadow many other activities until satisfaction is obtained, especially among youth and young adults.

Finally, each human seeks to avoid the plain truth about one's eventual demise. Keeping busy with a vocation, hobbies, raising children and enjoying entertainment help to deflect any conscious awareness about the finality of each person's existence. But always lurking, knowledge of one's inevitable death acts as a great motivator to enjoy what time we have left in this astonishing world.

Egoism and altruism

Although there exist many different forms of egoism, the rationale for this life style calls for self-interest to trump the interests of other people; that in the struggle to survive among all other forms, the self must pursue those goals which are beneficial to it. Carried to its extreme, egoism fosters a misanthropic attitude, where others are merely stepping stones toward one person's or one group's success. Therefore, egoists eschew regulations and controls from outside sources, such as from governments or other social structures if these regulations in any way impede the path of self-aggrandizement. In the case of egoism, we see the absence of synergy and collectivity as well as the lack of compassion for other people.

We suspect that the recurring forms and processes are responsible for such behavior, bestowing rugged and inconsiderate personal traits that fight against controls of any kind. This is why rational social policies and laws are required to curb such rapacious behavior.

Of course, most of us will act in ways that are helpful to ourselves, which certainly makes sense. Acting in a manner that is detrimental to oneself is foolish. Each individual member of the population wants the best results for himself or herself and cannot be condemned for such behavior. What can be criticized is the manner in which one goes about seeking self-improvement. In a synergistic civilization, acting according to the principles of adequate wisdom, we all exist as members of a greater whole and therefore are responsible for the success of all its members. Actions that benefit one person to the detriment of others are simply counter-productive and harmful.

Institutionalized egoism, such as totalitarian regimes, suppresses the will of its citizens, shutting down their self-interest in favor of the sole interest of the regime in power.

Egoism manifests itself in many corporate structures, where the overriding goals may work against the population as a whole. The quest for profits can often supersede the good of the entire population. Any institution that rebuffs or hides from constructive criticism acts out of self-interest and exhibits little or no social responsibility. This self-interest appears in many other social structures as well (including governmental institutions) and requires policies and regulations that promote activities to benefit all citizens.

Whereas institutionalized and personal egoism can be found throughout all of human history, institutionalized altruism has had many speed bumps, and even in this current civilization, human rights and protections are not universally practiced, which means that altruistic acts remain in the hands of good people who help others often without the support of national policies and regulations.

Whether or not altruism springs from forms and processes that produce such a trait, it remains one of the greatest qualities of humankind, since by its very nature, self-interest is superseded by the urge to help others whose circumstances have placed them in precarious and penurious situations.

Of the four standards by which adequate wisdom guides us—pleasure and responsibility, strength and compassion—altruism represents compassion, which in this writer's estimation is the greatest human virtue. Altruism also can represent the pleasure that one derives from helping others. Altruism can also exhibit the responsibility that all of us have to care for one another. And altruism requires great strength of conviction to toil among somber and poor conditions in any country on Earth.

Many religious and secular organizations promote altruistic acts. Oftentimes, governments and social organizations promote such behavior, like the Peace Corps or Doctors Without Borders, for example.

Could there be a more striking polar opposite than egoism and altruism? The egoist represents the rugged individual stepping over people to get ahead while the altruist tries to lift up the people whose circumstances have battered them down.

The manipulation of nature

Whether breeding plants, fruits, vegetables; breeding domesticated animals like dogs and cats; channeling the electron to create electricity and computers; or creating new molecules and new species, humankind has shown its remarkable ability to manipulate the natural world and create an incalculable number of inventions, products and gadgets.

Even the most cynical among us would have to admit that we have come a long way since caveman days! From manipulating fire and creating the wheel, we have reached a highly evolved state of science and technology that produces complex computer and software systems; space shuttles and space stations; Moon landings; CAT and MRI scans; artificial organs; cures for diseases; photovoltaic cells to convert sunlight into electricity; bio-engineered species; precision tools; highly entertaining telecommunications devices; and countless other wonders coming from the free will and design features of the human brain; including beautiful works of art, excellent musical compositions, clever works of literature, films that excite and thrill us and all the other creations that humankind has utilized in the natural world.

Despite the pettiness and struggles of humanity, we must congratulate ourselves! We have produced a civilization filled with creations and devices that make for a productive and busy biosphere and a fulfilling lifetime.

However, there is one form of the manipulation of nature that our ancestors would have never dreamed of, and that is the actual destructive manipulation of our planet's atmosphere and hydrosphere. It is quite evident that because of our pollution of the environment, we are actually altering the climate of the Earth, with potentially horrific consequences looming. In the 21st century, the supposed age of science and technology, we face people in powerful positions denying scientific truths and placing roadblocks to prevent intelligent management of the biosphere, often times simply because of short-term economic gains. What a terrible tragedy to witness this ignorance and avarice!

The brain, civilization and the universe

The common threads connecting the brain, civilization and the universe are synergy and individualism, the remarkable interconnectedness of countless parts all interacting to form greater wholes, while also performing very specific, individualized functions. The systems and sub-systems in the brain form an integrated whole that performs countless voluntary and involuntary actions to allow for movement, temperature control, appetite, habit, emotion, language, thought, volition and numerous other functions that make us robust living beings.

The communication among all brain parts is carried out by trillions of signals that provide information to the cortex and other brain parts. Each part has its specific set of neuron complexes to carry out specific duties, yet each part contributes to the holistic nature of this amazing organ.

Civilization consists of countless parts, each performing a particular function, and yet each part contributing to the entire social structure of human interactions. In the global economy, a problem in one nation now affects all other nations. Telecommunications systems, much like dendrites and axons in the brain, connect us to all parts of the world. International alliances (such as the European Union and the United Nations) attempt to settle disputes and provide worldwide or regional regulations for the betterment of humanity, aspiring to become the "cortex" for all humans. At this point in time, there remain various degrees of opposition to a universal command center, as many nations refuse to become subservient to a greater whole. If synergy has any say in the matter, one would expect the eventual establishment of a world brain or command center, so to speak, that looks out for all people and sets universal policies for the flourishment of the species and the planet.

As for the universe, it too has its individual parts and wholes, but these are perceived as physical forms without apparent meaning. We do not know why tiny strings or particles exist, but we do know that they form incredible structures, such as the atom, molecule, star, galaxy, planet and our very lives. We do see that these physical forms and processes act individually as well as combine to form greater wholes, all of which have contributed to the improbable evolution of living things.

Quantum mechanics seems to suggest that the universe itself is an interconnected whole, pulsating and vibrating, and somehow acting as an integrated system. Is it possible that the entire universe acts as a command center or brain with some kind of purpose beyond our understanding? Certainly this idea is speculative, yet somehow there may be a commonality among brain, civilization and universe.

Intelligence and wisdom

There are certainly many levels of intelligence among humankind, usually measured by an intelligence quotient, which points to the varieties of brain circuitries that endow each person with special qualities, much of which is passed along genetically from the parent sex cells. The inherited abilities or disabilities range from genius to severe mental retardation, the latter a result of dysfunctional brain processes.

Absent any injurious chance events during fertilization, intelligent parents will usually produce intelligent offspring. This intelligence encompasses a wide variety of skills, including excellent memory, superior learning and communication abilities and often the tendency for advanced abstract thought.

We are not certain what directs each person to excel in different skill sets, such as science and technology, medicine, mathematics, philosophy or the social sciences; or the ability for articulation, either verbally or compositionally.

IQ tests do not necessarily determine a person's musical or artistic abilities, but these are often passed along as inclinations or proclivities from the parents. But even in some instances, other processes in the brain may endow individuals whose parents have no artistic abilities to exhibit such skills.

Intelligent people do not automatically possess wisdom, as this quality requires a comprehensive view of the world. Most intelligent persons become superb practitioners of their specific skill sets and thus often become experts in their narrow fields of endeavor. Of course, occasionally there are people—we call them Renaissance men or women—who excel in a variety of fields of study and who do indeed concentrate on a wide world view.

As indicated throughout these essays, the search for adequate wisdom requires an overview of the forms and processes, both working interdependently, which are responsible for existence as we know it.

Thus intelligence alone, while being a wonderful gift of chance, circumstance and free will, does not, ipso facto, bestow wisdom upon an individual.

Design of Laws

Humans have created a wide range of laws which govern much of their activity, from nations to groups to individuals. Law making is found not only in the secular divisions of society, but in religions as well. Laws are designed and implemented until modified or changed, and usually operate as deterministic controls over the persons or parties concerned. Some religious laws or edicts are sacrosanct and therefore subject to little or no change.

Laws are often imposed to curtail or prevent humans from engaging in a wide range of harmful or questionable activities, but also to referee disputes among people or corporations or nations. Laws ought to be designed and implemented to enforce policies that promote an enlightened and humanistic agenda based upon the principles of adequate wisdom.

Laws vary from nation to nation and from state to state or province to province. Most countries have established a hierarchy of courts of law to examine the viability of existing laws and to make changes when deemed appropriate or necessary.

By observing the following categories of law, one can easily grasp the range of human activities governed by rules and regulations.

Air and aviation; antitrust; arbitration; bankruptcy; business and commerce; child rights; civil law; civil rights; communications; constitution; consumer transactions; contracts; criminal; disability; domestic (children, marriage, divorce, etc.); discrimination; immigration; entertainment; environment; estates and trusts; ethics; health; insurance; labor and employment; land; litigation; malpractice; maritime; media; mental health; military; natural resources; patents, trademarks and copyrights; personal injury;

property; public utilities; real estate; science and technology; sports; taxation; torts; and wills.

In some political alliances, such as the European Community, laws for specific matters affect all of the member nations. In most instances, laws serve as determinants of human behavior and need to be fashioned according to the principles of adequate wisdom.

The International Criminal Court (ICC), created in 2002, has the authority to prosecute individuals or nations that commit war crimes and crimes against humanity. Most nations of the world grant it authority to conduct such prosecution, with some notable exceptions, including the United States and Israel. It exercises its authority only when individual nations do not choose to investigate or prosecute alleged crimes in or by national states.

The ICC is a powerful example of the synergistic trend exhibited by the nations of the world.

War and hostility

One rather unfortunate result from recurring forms and processes is the continuous presence of wars and hostilities throughout the history of humankind. These acts of aggression can exist as simple hostilities among family members, neighbors and localities or can balloon into gigantic and fierce conflicts within nations (civil wars) or among many nations of the world community.

Misunderstanding, misinformation or downright belligerence often lead to vicious actions, which include the use of all kinds of weaponry; from spears, knives, machetes and guns, to poison gas, missiles, tanks and aircraft, in the attempt to vanquish an opponent. Some nations now use remote control drones to attack and kill their perceived enemies, while oftentimes killing innocents. Cyber warfare is the newest form of hostility designed by humans.

Wars have been fought for the acquisition of new land, for economic dominance, for the spread or suppression of religious beliefs, for the correction of what is perceived as a wrongful act against a nation or race, and a host of other reasons based mostly upon revenge, fear, hate, rage and mental pathologies.

The recurrence of aggressive personality traits as well as unconscious hatred and fear contribute to acts of war and hostility, especially when these traits are manifested in the leaders of countries, tribes or terrorist groups.

Even in everyday transactions among ordinary humans, there exists a range of perceived slights, insults and misunderstandings that can lead to verbal abuse or outright revenge. Hostility against minorities is commonplace.

Initially, the power of emotion often trumps rationality, leading to sustained conflict. The rational part of humanity eventually seeks to end some of these conflicts by the establishment of treaties and other resolutions. Military alliances can often end a conflict by conquering an aggressor, but resentment smolders and can lead to future conflicts. Some nations are so bent upon revenge and continuous hatred for other nations and their people, that no rational settlements can be reached.

In the immediate future, it seems quite unlikely for wars and hostilities to end, as they are part of the disparate human landscape influenced by recurring forms and processes. Perhaps one day, a synergistic world community might be able to settle disputes before the acting out of individual, group or national aggression. But in the meantime, we can be certain not only of death and taxes but wars and hostilities as well.

Bullying is a form of aggression and hostility that appears commonplace, especially among students. Bullies are those with belligerent and insecure personality traits, as well as those who are the products of ignorant or prejudiced families. These circumstances produce the ridicule and humiliation of other students who may be different for a variety of reasons, such as ethnicity, sexual orientation and the like. School authorities have performed poorly in protecting vulnerable students and need to do a much better job in the future.

Compassion and punishment

Using the guidelines of adequate wisdom, we know that we cannot arbitrarily state that each person or group is completely responsible for actions taken or policies expressed. We know this because in addition to free will, there are the persistent other factors (chance and determinism) that influence the individual or group. Thus, the dispensing of punishment, ostracism, disparaging remarks and other negative reactions must be tempered. Of course, we must remove from the general population those whose synergistic wholes are ridden with deleterious processes. We do this by imprisonment or in very few cases by execution, but in some instances capital punishment has been shown to execute innocent people. In a humanistic civilization, no death penalty would exist.

The point to consider is the treatment of those whose character is or has been harmful to others. If we assume that these criminal and antisocial types arise not only from free will, but from amoral determinants as well, then our treatment of them ought to be civil. They deserve to be treated as victims of their circumstances, as well as premeditators exercising their free will.

It's difficult to refrain from outrage when someone is killed, badly beaten or mercilessly bullied. The common reaction is revenge, then punishment. While it certainly provides some solace to the aggrieved victims, seeking revenge seems contrary to adequate wisdom. The problem, of course, is that the very nature of revenge is also a result of all determinants, including free will, chance and determinism. We face circularity here, as all things are interdependent, so that a vengeful person is not completely responsible for his or her actions.

Moral systems view events as black and white, as good events and bad events. People praise the good events and condemn the bad events. Then there are events having nothing to do with humanity. These amoral events are based upon chance and determinism and are beyond the scope of human interaction and therefore moral judgments remain inapplicable.

We might suggest another moral system, a holistic morality, which tempers the polar opposites of good and bad, and which softens moral outrages and tirades.

Dealing with the synergistic wholes and their interdependence and realizing that free will is not the only factor to be used in devising a moral system, we begin to approach an adequate wisdom that favors a moderate approach to bad and evil events. From racial slurs and ignorant comments to severe criminality, we need to be cognizant of the fact that these bad actions are not all based solely upon free will, but on all determinants.

When people's actions are not vicious but tend to be offensive to others and the environment, there's always the use of feedback, which often times corrects the bad behavior. Feedback is fine when it is not used as a hammer over someone so that the person issuing the feedback is not trying to manipulate someone for his or her purposes.

From this discussion we have reached the conclusion that consistently bad people need to be removed from the population but should also be treated with civility since not all of their actions can be completely attributed to free will. People often scoff at the use of a poor and disruptive environment when defending a criminal. Outraged citizens cannot find compassion because they believe the criminal to be totally responsible for his actions. We know that this is not the case. Adequate wisdom suggests the application of compassion and fair confinement, since all different forms and processes comingle in the lifetime of every single human being, group or civilization, leading at times to antisocial behavior.

There is also the matter of redemption, whereby society grants a reprieve to one who has not committed a capital crime and thus, for example, a first time offender is generally allotted a lesser punishment with the goal of rehabilitation. Unfortunately, in many cases, we are faced with repeat offenders whose character has been permanently scarred by chance and determinism.

Most of us are fortunate to have been molded by benign determinants and are equipped to use our free will for the good of humanity. Unfortunately, there will always exist a small number of damaged members of the species.

Education of young people

Educational systems vary across the globe, but most have one thing in common and that is providing children with basic skills to read, write, think and learn about the history of humankind, as well as the biological and physical universe.

After reading, writing and a basic vocabulary are instilled into young minds, there is often the mad rush to hammer children with a variegated assortment of dates, facts and figures, without first setting out the broad view of existence. Children ought to be shown a simplified version of the major parts of the universe and how they fit together. Then details can be taught.

Since some lesser-skilled students will not do well in certain subjects—for example, algebra, trigonometry, physics or a foreign language—they ought to be given a pass as long as they learn reading, writing, vocabulary and thinking skills to navigate their way through life. Many academically challenged children experience so much frustration from mathematics or foreign languages, for example, that this frustration leads to a resignation from the study of other subjects that could be appealing and useful to them. Of course, all children should be exposed to all major subjects, including the difficult ones, but there comes a point when forcing such knowledge down a child's throat becomes counter-productive.

Holding back students from graduating because of poor accomplishment in subjects they will not need as adults undermines the entire system of education. Each student is equipped with certain abilities by chance, determinism and free will. For those with high level ability, of course math, science and foreign languages can be handled by them and do not create an unnecessary amount of frustration in learning about them. But for students whose skill sets are below average, it is simply unfair to force a predetermined list of subjects into their minds, many of which will only cause despair and trigger a large increase in the drop-out rate. Course design should be tailored for each individual's abilities.

In addition, school administrators must implement sensitivity training to promote compassion for all types of students to prevent bullying and belittlement of those who appear or act differently from others.

Questions for teenagers

Here are some questions that teenagers could ask them-selves. By asking and answering some of these questions, teenagers may gain a foothold into the creation of a flexible and rational life plan. They may be able to see themselves as both an individual and as a member of synergistic wholes, such as family, school, mates and members of social, ath-letic and entertainment groups. By pursuing pleasure and responsibility and by exercising strength and compassion, they have an excellent chance to prosper as they evolve into mature adults.

1. Who am I?

2. What do I think of myself?

3. How do I fit into the world?

4. What kinds of personality traits do I exhibit?

5. What talents do I appear to possess?

6. How do I relate to my family, friends and others?

7. Do I spend too much time thinking about my appearance?

8. What are the beautiful things in life and the universe?

9. What are some ugly or disturbing parts of the world?

10. Do I see the big picture that exists beyond my own life, friends and family?

11. Why are some people much worse off than I? Or, much better off?

12. How do I understand puberty?

13. How do I cope with sexual feelings?

14. Does the need for sexual contact affect the way I dress and act?

15. Am I able to discuss sex with my parents?

16. What's the purpose of education and other forms of socialization? Am I simply going through the motions of studying without any real grasp of the subjects? Do I realize that education is my ticket to success?

17. Am I being influenced by my peer groups to take actions I believe may be wrong? Should I search out different people with whom to associate?

18. Do I make fun of people different from me?

19. How much time, if any, do I spend with close friends discussing the nature of existence—of life, religion, death, good, evil and other such issues?

20. Am I totally occupied by the telecommunications and electronic age? Do I constantly text, tweet, e-mail or use online services to connect to other people? Do I watch too much television or play too many video games?

21. Do I ever allow myself time away from friends, electronic gadgets and other games and diversions to engage in a quiet period of thought?

22. Do I act without thinking about the consequences?

23. Am I very critical of myself?

24. Do I always find faults in others?

25. Do I make hasty decisions about people and ideas or do I try to look at many sides of an issue?

26. Do I question the political and religious values of my family and friends? Do I automatically follow

religious and political beliefs of my parents and friends, or do I try to develop my own religious and political views?

27. Am I reading newspapers, magazines, internet sites, books or e-books?

28. Am I expanding my knowledge through school as well as through my own initiatives?

29. Do I stand out as an independent member of my family and yet partake in the synergy of the entire family by cooperating with family members?

30. Can I take criticism without automatically viewing it as an intrusion into my life style? Or can I use feedback from others to make adjustments in my actions which will eventually benefit me?

31. Am I compassionate about people less fortunate than I? If so, what am I doing to make life better for others?

32. Am I more important than anyone else?

33. Do I have self-control to handle urges and fantasies that may be harmful to me and others?

34. Do I balance pleasure and responsibility?

35. Do I show strength and compassion?

Individual humans

The key frame of reference for each person is his or her awareness of the world. Even though each human has basically the same physiological structures, each brain is wired differently and each person looks and acts differently from all others, excluding identical twins. And even these identical twins will exhibit slightly different personality traits. The main point to be made is that it is only through each individual mind that the world comes into focus.

The influences upon the mind and personality, from within and without, lead to a wide range of behaviors, from egomaniacal to altruistic, from benevolent to treacherous. Because of chance, determinism, free will and design, the individual human being unfolds into itself because of the assemblage of forms, processes and ideas, creating a unique synergistic whole.

Whether one leads a life of leisure or despair, wealth or poverty, rationality or irrationality, kindness or cruelty depends upon the myriad forms and processes interacting with one another. Through history up to the present day, there have always been economic disparities—divisions or strata among all people, so that there have always been the poor, the middle class and upper class; there have always been the uneducated and the educated, including a stratum of very bright and extremely creative people who can arise out of any economic class.

There have always been people who for a variety of reasons seek power and control over others, who have little or no regard for the feelings of others and who will do anything to achieve what they want and desire. On the other hand there always have been people with great compassion, love and sympathy for others. Character and personality traits, along with circumstances, will define which type of lifetime one follows.

There are all types of characters among the seven billion human beings, those with sparkling personalities who are alive with vibrancy and creativity and humor; while there exist selfish persons who only seek self-aggrandizement. According to adequate wisdom, this situation will not change.

Human eccentricity

Action or thought that appears outside of the normal range of behavior exhibited by most humans is often called eccentric. A person may be rebelling against existing norms and thus will dress or act in a strange or unusual manner. Sometimes a person adorns himself or herself with permanent markings on the body or with one or more piercings. Eccentric people might behave irrationally. Some may hold views that are far beyond the mainstream of current thought.

What could account for such unusual behavior? Well, we know by now that in addition to free will or conscious choices, we stare right into the face of chance and circumstance that affect someone's mind. The conscious (and unconscious) mind is certainly affected by unusual brain chemistry as well as uncommon personality traits.

Depending upon the degree and injuriousness of the eccentric behavior, social structures may quell such behavior by various means, including ostracism, critical feedback or legal action.

Throughout history there have been a fairly large number people who were judged eccentric at the time but whose views or actions are now acclaimed and honored. Before any actions are taken against such people, members of the population ought to be leery of casting aspersions against odd or unusual behavior and thought.

In these days when the population concerns itself with possible acts of terrorism, eccentric people might set off a red flag and have their behavior watched or cataloged.

But, in the broad view of this behavior, eccentric people often bring forth new creations and ideas that contribute to civilization; so their reputation should not be automatically tarnished.

There are also many eccentric or radical groups of humans sharing the same point of view. Each one has different purposes and goals, some of which may be very harmful to the population and therefore warrant observation or elimination.

Love, attraction and friendship

Throughout the ages, love and the concept of love have permeated civilization. Love means different things to different people, but most of us agree that it comprises the deep well of emotions (including sensuality), personality traits and character.

As a verb, it signifies one's strong feelings for people or things and events. "I love her. I love him. I love my child. I love my friends. I love myself. I love to go water skiing. I love my job. I love the way she plays the cello."

As a noun, it symbolizes tenderness, care and concern for other people, as well as deeply felt attractions.

As an adjective, it simply means pleasant and enjoyable, such as a lovely day or a lovely dress.

The most common form of love is that found between two people, who are, for example, lovers or those engaged or married. It is often stated that a good chemistry exists between the two (maybe partially attributable to pheromones?). If we examine these so-called love relationships, we often find that physical and sexual attractions, as well as personality traits, first bring people together. There are exceptions, of course, but love takes time to grow. In a longterm relationship the love that two people experience is a joyful (lovely) event.

The more a relationship is based upon physicality, the more likely it will fail. The more a relationship is based upon mutual respect and tenderness, the more likely it will succeed. The love that a person has for another can turn dramatically to hatred, especially during a nasty break-up, separation or divorce. Love can also be short-lived, especially among young people who cannot distinguish the difference between infatuation and love.

It is exceedingly difficult to describe what love actually is for each different person, but it surely reflects the conscious and unconscious states and certain brain complexes. When sexuality and attraction are added to respect and care for another person, and these qualities persist, even in some cases for a lifetime together, we witness the actualization of love.

Unfortunately, because of chance, circumstance and free will, marriages falter and dissipate. In countries such as Sweden and the United States, where divorces are easily obtained, the success rate of a marriage is under fifty percent. In countries where divorces are virtually impossible to obtain or shunned by cultural forces, such as India and Japan, the divorce rate is under two percent!

There are many marriages where love and sexuality have been greatly attenuated, and yet for the sake of children or other needs, the marriages remain intact.

There is also the phenomenon of self-love or narcissism. Loving oneself is healthy and productive, assuming, of course, that one also loves others as well. When one is totally absorbed in oneself to the exclusion of all other people and things, then there exist deep psychological problems caused no doubt by chance and circumstance creating highly self-centered traits.

As far as attraction is concerned, our personality traits, emotions and aesthetic endowment lead us to seek out that which pleases us. Based upon our makeup, we are attracted to music, art, conversation, film, literature, games, sports, and an entire range of people and things abounding in the world.

There is also the idealized love for all humanity, all creatures and the entire biosphere. This kind of all encompassing love is not as common as we would like, since the dissonant personality traits found in some humans work toward a pessimistic or misanthropic view of the world.

Finally, there is a form of love that does not involve sexuality. It is friendship, one of life's most wonderful forms and events. Friends share good and bad times together. They form a link that is difficult to break. Good friends are a very special treat of existence and make life much more bearable for many members of our population.

Love, friendship and creativity are three hallmarks of existence!

Individual culpability

Is each human responsible for his or her actions? This question is of paramount importance when considering the assignment of blame. In today's moral climate, many believe that each one of us is responsible for what we do and say. If we consider that free will allows each one person to act according to a premeditated thought pattern, then we would agree that these actions, bad or good, can be criticized or rewarded. But if we recognize that along with free will comes chance and determinism, then it might be argued that we are not fully responsible for all our behavior, that instead we are sometimes acting according to influences over which we have little or no control.

The plight of each person is indeed determined by circumstances, by the whole of existence, and not just an individual action but by a series of multiple events, both past and present. One cannot dismiss the wholes that make up existence. The atom, the star, the planet, the cell, the brain, the galaxy, the civilization—all these components operate interdependently and bring forth a constantly changing and yet paradoxically steadfast reality. Forms are structures that exhibit steadfastness, while processes are events that bring about change.

Thus the individual human cannot be completely culpable for actions and behaviors that exist because of the random and deterministic confluence of forms and processes. However, since free will does exist, volition can cause bad acts to occur, and thus the individual can be held culpable for such actions.

The problem of free will and human design versus the other determinants of behavior is that one can use the amoral determinants as an excuse for certain behaviors. It is a thorny issue yet to be resolved.

The wise person must be able to draw the distinction between what is truly based upon free will from what is based upon chance and determinism. Of course this is a very difficult task, and thus we must be very careful when castigating people whose actions may result from circumstance other than free will.

Group and institutional culpability

Those with common interests and goals will form groups whose strength increases because of the synergistic combination of the individuals within the group. These groups, assembled for good or bad purposes, share common forms and processes. The question arises whether these groups are culpable for their actions or whether they act in accordance with chance and determinism and remain amoral.

There exists an enormous number of groups, associations, clubs and the like, some of which have explicit views on morality, religion and politics. The vast majority of these groups do not appear to use a broad-based philosophy that incorporates the principles of an adequate wisdom. They become so single-minded in purpose that, in some cases, they ridicule other individuals or other groups, institutions and nations.

As any closed-knit group, they resist criticisms. Insofar as each member coalesces with others with the same viewpoints, we know that the circumstances which brought them together include free will but, again, they were and are subject to the forces of chance and determinism beyond their control. Are they culpable for their statements and actions? Once again we face the dichotomy of distinguishing between their free will and other forces beyond their control.

Since the formation of social groups has always existed, we must understand that other forms and processes helped to create them. In that sense, they may not be entirely responsible for their actions. Nonetheless, if these groups violate any of the tenets of adequate wisdom—such as lacking compassion and responsibility—then they are culpable and must be altered or disbanded so that the flourishment of humanity is not jeopardized.

As for major institutions, like corporations and governments, the same logic applies. If their actions harm others or if their policies work against the flourishment of humanity, then they, too, should be modified or changed.

The emergence of human stars

One fascinating component of existence is the continual emergence of bright spots of life from individuals and groups who provide the human race with a broad sweeping range of talent and ability that adds greatly to the cultural pleasure chest. Even those whose main talent is a sparkling and refreshing personality bring joy, comfort and laughter to others.

Surely all facets of existence have an influence in creating these special people and groups –from free will and determinism (heredity) to chance and synergy, all coalescing to form these special human jewels.

The drive to express oneself runs deep throughout human history, and we have been fortunate to reap the benefits of artists and artisans, singers, dancers, composers, musicians, writers and playwrights, filmmakers, conversationalists and all others who provide us with pleasure and food for thought.

Others expressing themselves and providing ideas and producing artifacts include inventors, scientists, mathematicians, educators, philosophers, doctors, engineers, researchers and so many others to whom we are grateful.

Imagine a human race without such human stars! What a dull and dreary lifetime we would lead.

(One wonders what is to become of all the art, music, literature, drama, philosophy, film and all other cultural and scientific contributions made by these human stars as we move into the future? We have already inherited several thousand years of culture and haven't really digested much of it. Imagine if the creative productions of the past several thousand years were digitized and available for our viewing! What about hundreds or thousands of years from now when everything is available for our viewing and listening pleasure? How can we possibly appreciate the enormous amount of human creativity produced over the forthcoming centuries? The only foreseeable means to appreciate even a small portion of the creative output by our species in the future is to experience much longer lifetimes, which now seems quite probable.)

Aesthetics and adequate wisdom

Judgments about food and drink, art, literature, music, theater, film and the beauty of living things constitute a very difficult subject matter upon which to comment. Most people believe beauty remains in the eye of the beholder, and that makes perfect sense, since we know that each person represents a specific character based upon personality, inherited tendencies and talents, as well as circumstances providing each person's lifetime with a unique view of the world.

Let us look at every day events that involve the central nervous system.

What we eat or drink may be judged by our sensory buds and brain to be quite excellent, mediocre, distasteful or rotten. Some foods and drinks may take a while for the person to acquire a taste for them, so that the initial negative reaction turns to a favorable reaction because the person has learned to enjoy the taste. In another words, repetition can alter the initial aesthetic judgment.

But then there's a psychological element to what we eat. In a few nations or communities around the world, it is not uncommon for citizens to eat unusual foods like fried black scorpions, lamb's brains or jelly fish. It is safe to say that these delicacies would not be found in the diets of many people of the Western world, but one could conceive of learning to eat and enjoy them through repetition. However, the emotional and psychological reaction would most likely steer some people away from these dishes. So it seems that we can prejudge and aesthetically disapprove of objects whose names alone or whose appearance or taste we believe would not be pleasant to our palate.

Moving on to the appreciation and aesthetics of music, we find an entire spectrum of likes and dislikes based usually upon the social milieu and age of the individual listener. Like food, through repetition it may be argued that almost anyone can learn to appreciate most musical genres and compositions. But as often the case, people tend to pre-judge what kind of music they like, beyond that which they already enjoy. Popular music tends to reflect the attitude (lyrics), the rhythm of the times, and the peer groups to which they belong. Blues, jazz, country music, show music, hip-hop and rap, alternative, pop and rock seem to dominate the musical sphere for most people in the current civilization. The songs are generally short in length and easy to grasp. Repetition of these songs adds to their popularity and allows for most members of the population to sing along with them or dance to them. Thus, this type of music is aesthetically pleasing.

When it comes to more difficult music, such as classical music and opera, there is a general prejudgment of these genres. Exposure to classical music is not common, and most people will not learn to like it as they are not willing to listen to a classical piece once, let alone the several times that they must listen to understand and appreciate the entire work.

Can we state that one piece of music is better than another? Could we compare *(I Can't Get No) Satisfaction* by the Rolling Stones to *Carmina Burana* by Carl Orff? We cannot say that one piece is better than the other, only that the two compositions are different from each other. We could say that the former is several minutes in duration while the latter is nearly an hour long. We could say that the instrumentation and harmonies of the latter are more intricate than the other; but in terms of aesthetics, we really cannot say that one is better than the other.

(Many hard core classical music lovers would most likely object to the comparison being made here and claim that the work by Orff is much better than the Rolling Stones song. But this highlights the problem with investing one-self in one genre and pledging complete loyalty to it, while denigrating other less intricate forms of music. The point is that each creation stands on its own and cannot be debased. The trap is set for both groups—each snubs the other's choices instead of acknowledging that each work is different from another rather than of lesser quality.)

The problem arises when each self denigrates an art form. It's one thing to say "I don't understand this piece of music" and another thing to say, "this piece of music stinks!"

Visual reactions to art and film also engender positive and negative reactions, but once again, it is difficult to say that one piece of art is better than another, or one particular film is better than another. Is a work by Claude Monet better than one by Marc Chagall? Is the film *Crimes and Misdemeanors* better than *Avatar*? Because of the tremendous differences in personal tastes, ages and personalities, it really is impossible to reach any definitive aesthetic conclusion.

When it comes to the aesthetics of literature and poetry, we deal with a multiplicity of themes, imagery, character development, metaphors, similes and a variety of other components that make the written word more than a sensory experience as words trigger our imagination. The neocortex is alive with interpretations and reflections from literature (as well as film, art and music.) Readers will find some works quite satisfying while others find them boring or of no interest to them. Personal background, education and taste all impact upon the judgment of a work of literature.

Finally, the most difficult aesthetic to discuss is that of the beauty of nature, especially the human form. There is not a single standard by which to judge the beauty of human beings because of the various cultural and physiological differences among peoples and races of the world. But within each race, there exist standards of beauty that people appear to employ; whether it is based upon age, musculature, facial features, hair, sexuality, personality and all the intangibles that comprise an aesthetically pleasing person.

Are we justified in claiming that one person is better looking than another? From each specific individual's viewpoint, we can answer, yes, that each self knows what it likes and what it dislikes.

As for the beauty of a human being, each person has his or her own sense of what is appealing and attractive. The tragedy occurs when a single person or group of people ridicule the outward form of another person and employ pejorative descriptions and name-calling.

The combination of good looks, innate intelligence and pleasant personalities are gifts from chance and determinism and most often bestow quite an advantage for those with these qualities in their travels through life.

Instantaneous and worldwide communications

Communications (including ideas) among our ancestors mostly occurred in the town or village in conversational form. Very few people in other regions or nations could be privy to their ideas, except through hand-written manuscripts and notes. Thanks to the invention of the printing press in the 1440s, books, pamphlets or even single sheet notices could travel farther than ever before. But it took fledgling postal systems mostly carrying personal information to grow into international mail delivery in the 17th century finally to circulate the written word to literate people around the globe.

With the manipulation of the electron and the electromagnetic spectrum, a telegraph service formed to spread news, then radio broadcasting arrived, followed by television—and in a split second of evolutionary time, humankind launched the Age of Communications, evolving into wireless, land- and satellite-based technologies. Now, each individual or group can become its own broadcasting station or network and reach out to the entire planet with instantaneous communications.

Still in its early stages, the Internet has blossomed into a most remarkable human concoction, with the globe literally alive with frequency transmissions that carry videos, music, conversations, encyclopedias, dictionaries; and websites for a really unlimited number of human preoccupations, such as clothing, sex, fashion, beauty, sports, films, music, cooking, vitamins and health and financial reports.

This 21st century phenomenon results from the synergies of countless entrepreneurs, the application of the electron and technology, human design, chance and circumstance. The communication and electronic bonanza has created the beginnings of a worldwide brain, so to speak, with each participant acting as a neuron by spreading information ultra-quickly and throughout the entire form. From afar, it seems as if an invisible hand is helping us to create a new synergy—a collectivity of information and ideas shared either by one person or by the entire human world, whose final form is not yet recognizable.

What is the influence of sexuality?

The process of biological evolution drives the replication of species and rewards human beings with enormous pleasure when they engage in and consummate a sexual act or union. There are very few who would deny the powerful feeling of ecstasy when reaching orgasm. Apart from those individuals who have chosen to abstain from sexual and other physical pleasures, the vast majority of people are deeply motivated by the sexual drive. A combination of a satisfying sexual union and a harmonious social union between two people often yields a longterm, loving relationship.

Very few societies celebrate the orgasm. It remains a private, personal phenomenon in Western civilization. Nonetheless, the sexual drive remains one of life's dominant influences. It often provides a source of ecstasy, puzzlement, release, dismay, frustration, jealousy and confusion among many of our population. The advent of sexuality occurs in young girls and boys when their minds are not fully developed and thus are subject to curiosity, doubt and problems dealing with such enormous hormonal changes within the body.

We know that there exists a very wide range of personality traits and emotional states that experience sexuality in many different ways. For some members of the species, sexuality can be an all-consuming enterprise, which deeply affects their daily lives until satisfaction is reached. One can never underestimate the power of sexuality, which influences not only the mind-set of each person but also the manner in which one acts, dresses and presents oneself to others.

The notion that sexual frustrations are sublimated, that is, are replaced by the preoccupation with music, literature, design or any other non-sexual creative activities still has great merit.

The personalities of girls and boys are exhibited early in their childhood, so that different personality traits will lead a youth to handle sexuality in many different ways. For example, the extrovert will more than likely be open about sexuality than would the introvert. Most teenagers are understandably oblivious to the quest of searching for the nature of existence, since their frame of reference is based mostly on sexuality, popularity, appearance, approval and performance in sport, the arts and scholarship.

So intensely personal is the sexuality of people that many youths and adults cannot speak frankly about their feelings and drives. Repression of sexual urges is not uncommon. Most religions have a direct influence in repressing the sexual drive of humanity. But, as part of the social contract, curbs on sexual appetites are useful when these threaten others.

A discussion about or a display of the sexual organs is taboo in polite society. There remains an inherent shyness in matters to do with sexuality when in fact the sexual organs are part and parcel of the human landscape and ought to be open to discussion and celebration.

There is a widespread phenomenon of using crass remarks about people's appearance, including their sexual organs. Sexual humor mixes at times with ignorance and malice to create public humiliation of people. This dark feature of humanity is not likely ever to disappear.

Adult Pornography

If we are to look at major forms and processes and agree that the sex drive is a powerful part of human existence, then we cannot exclude a discussion of adult pornography, which is a pervasive phenomenon in a great majority of nations, with a tremendous number of websites offering all kinds of sexual offerings. Of course, pornography has been prevalent throughout the ages, whether found on cave drawings, on canvas or in books, theaters, films, magazines and the internet.

We discussed the tremendous influence that sexuality has among much of the population. In the so-called proper society, adult pornography is regarded as crude and distasteful, and admittedly much of it can be extremely offensive to one's sensibilities.

But we must also admit that the strong sexual urges and fantasies of many people fuel the spread of pornographic materials and make this off-color activity a rather profitable venture. Adult pornography is usually protected by First Amendment rights in the United States and by similar protections in many nations throughout the world.

Because all Western religions regard pornography as sinful, there is an ongoing tension leading to continual religious outrage and preaching against it.

Using the guidelines of adequate wisdom, we must recognize that the varieties of sexual expression and events are the result of the interplay among forms and processes, of the recurring biological, social and physical influences. The opprobrium cast against adult pornography assumes it does no good and only harms members of the population. However, some brave psychologists and sex therapists might argue that sexual fantasies, if not appeased, may lead one to severely uncomfortable levels of frustration. In some instances, adult pornography might actually benefit society by providing a release for those who are sexually frustrated or for those who might otherwise commit sexual crimes.

Sexuality remains one of life's most mysterious and intriguing components, and we can never really quantify the imagination and fantasy life of each human being. Suffice it to say that sexual acts provide pleasure and happiness for our population. They replenish the species. They offer a spice to a life that might be otherwise dull and dreary for non-achievers. But, of course, the powerful sexual drive, without self-control, can and does overwhelm many individuals and produces a wide variety of actions that might be harmful to themselves and others.

For child pornographers and pedophiles, a brief discussion of them seems warranted, as they stand out as heinous phenomena in most civilizations. We know that sexual fantasies and the urge to have sexual contact or filmed or photographed sessions with underage youth exist throughout all populations. Young people must be protected from predators. We recognize that determinism and chance play a strong role in the behavior of predators and child pornographers. As such, the incarceration of these predators is a primary concern for society, but once again we should take into account the role of chance and determinism. These predators are subject to overpowering sexual urges bestowed upon them and they have neither the ability nor the will power to forge self-control. They have lost grip with civilization and common sense. The consummation of a forbidden sexual fantasy is also propelled along because the predator realizes that life is finite and that unless acted upon these powerful urges will never be realized.

Just like everyday criminals, sexual predators act out according to flaws in their makeup; and the incarceration of them, like any criminal, does not tend to change their urges, because, as stated frequently throughout these essays, most personal and biological inclinations and predispositions are given and remain somewhat fixed during any lifetime. (Check the recidivism rates for proof of this statement.)

Orgasm and evolution

How smart is biological evolution?

Orgasm, that's how smart.

Without orgasm in men and women, there would be a planet virtually barren of humans. Just like bees pollinating the flora, so do humans achieving orgasms keep the Earth populated.

It might have taken billions of years for orgasm to evolve, but most of us would agree that it was certainly worth the wait.

Thanks to the limbic system of the brain ordering the release of endorphins and involuntary contractions around the muscles surrounding the sexual organs, the ecstasy reached when orgasm occurs creates a unique euphoria throughout the body. There is no greater physical excitation throughout humanity in such a short time period.

Longer term euphoria of less physical intensity occurs in many facets of life, including a loving relationship, the creation of something new, the accomplishment of competing in sports, entertainment, business, academia and so on.

Thus evolution and determinism reward us for reproducing. And free will and human design reward us by creating a life worth living.

Self control of emotions and sexuality

As we have seen, the self is constantly interacting with other beings and other forms and processes. The self may find itself entangled in arguments, animated discussions and personality clashes where it loses control of rationality and strikes out in an uncontrollable fashion, most often to its own detriment.

Control often manifests itself as a personality-driven trait. As we know, there are wide varieties of personalities that handle stress or jealousy or irrationality in different manners.

We would be robots if we did not occasionally lose control and show a bit of anger or frustration. Each human has a different boiling point beyond which control is lost.

Obviously, the persistent loss of control creates harmful stress to the body. A person often losing control finds fault in most things he or she does or in what others do. Anger swells up quickly and often, and it is obviously best to avoid such a person if possible.

More often than not, self control grows with age and character development. By nature, a young person fights against outside influence and thus may be understandably given to periods when self control is lost.

Because of the wide range of influences from chance, determinism and the emotional state of the unconscious, self control will always vary greatly among all people.

One particular area of concern is self control of the primal sexual urges that pulsate among most members of the population. The conscious and unconscious mind along with hormones and sexual organs can create powerfully sweeping sexual urges that sometimes overwhelm the self and present formidable challenges to the control of sexual fantasies and desires.

The extent and power of these urges depends upon chance and determinism, and the control of them depends upon the strength of one's will power. Even with social mores and laws prohibiting the consummation of questionable or illegal sexual acts, the self is often bewildered, confounded and torn by the need to achieve its sexual goals as opposed to the consequences of such actions. No one should ever underestimate the enormous influence of sexual drives throughout our population.

We should mention that because of the very wide range of personality and emotional types, there are a great many human beings whose sex drives are not an all-consuming fascination and who are quite capable of controlling or consummating their urges with minimal difficulty.

Regrets

A common reaction to negative events of the recent or past history in the lifetime of a human being is the phenomenon of regret, that is, the feeling that one might have done a better job of performing a task, reacting to another human or groups of humans, or saying something differently or acting in a better manner.

People often second guess themselves when they have time to reflect on their words or actions—or their lack of action. These regrets occur in every walk of life and create lingering doubts about one's presentation in the world. Often times, making a mistake might lead one to regret that event. More often, regrets occur when time has passed and one is able to exclaim, "Why did I decline that job offer? Why didn't I tell that person I cared for him or her? Why did I treat my parents so poorly? Why didn't I tell the truth?" And so on.

It seems like a very natural trait to reflect upon or questions one's relations with others and the actions taken. Regrets can serve as constructive feedback, if past mistakes are used as a warning against present and future actions. However, the persistent preoccupation with the past and some unfortunate events that occurred can prevent a person from taking chances and acting assertively when necessary. Regrets can drag on a person's psyche.

Remember that chance and determinism are factors to contend with; a person's free will is modified and influenced by these modes of existence, as well as all other social and personal relations. Mistakes or "bad moves" ought to be downplayed for the most part. Regretting leads to further regretting and second guessing, with no end in sight. However, since character and personality traits drive much of human experience, those who live in the past will find much more room for regrets.

It's best, if possible, to continue forward with day-to-day activities and minimize or eliminate regrets. Of course, one should never forget the past—the good and the bad—as it does contribute to and influence the entire personal tableau. But constantly punishing oneself for past mistakes is self-defeating.

Humanity versus the rest of the animal kingdom

We humans have risen to the top of the intellectual ladder of evolution, post-dating mostly all other living forms and appearing to be the actual king of the kingdom. We know that there exist some rather intelligent and cunning creatures among us (a nod to dolphins and wolves, respectively), but none reaches the pinnacle of thought, conversation and creativity exhibited by the human race. We have domesticated several creatures, especially dogs and cats, which provide enormous companionship and pleasure; and we raise chickens, cattle and pigs for our eating enjoyment.

We are truly amazed by the complexity and super-abundance of all living creatures, but we do proclaim our superiority among them all. If certain living things (like insects) seem to annoy us, we kill them, either by stepping on them and crushing them, or by spraying noxious chemicals on them. Some of us hunt animals for sport with no regard or concern for their lives. Thankfully, some nations protect certain species from extinction.

Much of the animal world is oblivious to us. Most birds are skittish around us. Most marine life is alien to us. Those mammals more closely related to us can occasionally tolerate our presence.

The animal kingdom, although arranged in a hierarchy, is not one big happy family. Evolution and natural selection, among other processes, keep individual species together but for the most part separate from other species.

In addition, the carnivorous nature of evolution and natural selection where species kill and eat other species is a bit disheartening for those who feel compassion for all living things. Having tasted flesh, we find it very difficult to refrain from such activity; but to all vegetarians I salute you!

If there is an infinite number of universes, I can rest assured that at least one planet contains living things which do not kill other creatures for sport or food.

One interesting side bar is the belief held by a majority of the population that humans have souls and move on to an afterlife. What of the ant or spider or cow or elephant? Do these living creatures have a soul? Do animals move on to another dimension? If not, why would humans be able to move onward after death and not the entire animal and plant kingdoms?

Most answer this question by proclaiming that humans are God's chosen species, an explanation approaching the height of arrogance and stupidity!

The individual and the state

A major struggle throughout all human history has been and still is the role of the individual vis-à-vis the collectivity (group, institution or nation). Which wins out, so to speak? Is the nation always more important than the individual? Is the individual as important as or more important than the state?

If the patterns of synergy are correct, then each group, institution or state is greater than and different from the individuals who comprise it. But greater does not necessarily signify better than. So this suggests that there is a balance between each person and each state—a social contract that allows the greater whole to govern the lesser wholes (person, institution, group) and to nourish and protect them and promote their interests, as long as those interests do not conflict with those of the greater whole, assuming this greater whole is based upon rationality, freedom and democracy.

This balance has been and always will be a delicate, tenuous one. Historically, the state has reigned supreme over its subjects. Even in the democracies of ancient Greece and Rome, only privileged citizens had access to voting rights. In modern, humanistic democracies, the rights of individuals are protected and nurtured up to a point. The modern social contract requires each individual to follow myriad laws, regulations and customs in exchange for protection and services.

There are problems between individuals and the state, between institutions and the state, and between certain groups and the state. There are problems between individuals and institutions (e.g., a union member and his employer). These tensions are the result of forms and processes in conflict with one another and require mediation and arbitration and the issuance of laws or judgments to settle many issues.

One must always watch out for those rugged individuals who seek to undermine the role of government, for they do not comprehend the big picture of existence, wanting only to thrive on their own without care or concern for our entire population. Conversely, we must watch out for a government that tramples on individual rights and freedoms.

Absolute values should dominate the mind-set of any human and social whole, and that is self protection, pleasure, compassion, strength and wisdom. Common sense would dictate that good political-economic systems protect its citizens, its flora and fauna and its atmosphere, while not harming other people in their nation or in other nations.

Decision making

Until one has begun to understand the world and oneself, decisions may arise from mere instinct or the beck and call of others. Parents, extended family, friends and teachers will offer opinions, suggestions or regulations.

Instinctive decisions are based upon a variety of needs, such as hunger, personality traits and the need for approval. Adults and peers have a strong role in the lives of youngsters, even though it seems instinctive for young people to rebel against authority at times.

Before one can make informed decisions, common sense and personality traits tend to rule the day. A young person has the common sense not to jump out of a moving car. Yet, a personality trait that presents itself as a daredevil, upon being coaxed or not, may cause a youngster to make such a non-commonsensical choice.

As a general rule, informed decisions that are based upon wisdom of the world take time, but commonsense exhibited by young or old can lead to wise decision making.

Policy decisions based upon adequate wisdom

Because of the recurring nature of events and phenomena, we have said in these essays that certain forms and processes are inevitable, that we are not completely responsible for our actions. As such, one might be resigned to believe in a fate that is stronger than our will power to make positive changes.

However, it is the act of wise policy making, in the form of laws, regulations and ethics that can have a positive effect upon the human condition. Even though negative traits and prejudicial views will always continue throughout the present and future, humans are capable, through free will and adequate wisdom, to correct or at least mitigate the deleterious side of humanity.

From the Magna Carta to the United States' Bill of Rights to the United Nations peace keeping forces to the European Union's list of fundamental rights, and from a whole host of subsidiary laws and regulations, many, but certainly not all, human beings are the recipients of protections from some of the harmful events recurring throughout history.

Policy makers ought to encourage pleasure and responsibility and strength and compassion throughout the land. They must promote the flourishment of humanity and the protection of the biosphere, hydrosphere and atmosphere.

Policy makers must create opportunities for the poor, the uneducated and the physically and mentally disabled. Policy makers ought to discourage rugged individualism when it conflicts with the promotion of good deeds for all members of the population. If such policies are not carried out, then chance and determinism guide the evolution of all species in a haphazard fashion. Remember the idea of synergy—all things combine for a higher purpose. Self-seeking and self-promoting policies work against the fabric of existence.

Why are there changes in belief, attitude and opinion?

If personality and character are virtually fixed for each adult human being, if forms and processes recur and therefore create steadfastness, what creates the ongoing variability in people's feelings, actions and attitudes towards themselves and the world around them? It is the rational and creative part of humanity that encourages new ideas.

We change with new information and knowledge about the world. From the beginning of human thought through the present-day, opinions and attitudes about the Earth, the celestial sphere, the geosphere, the biosphere and humanity have changed because of discoveries and inventions that present new ideas and facts that are not disputable.

While determinism promotes a steady and unchanging routine, chance, human design and free will provide for changes over months, years or lifetimes for individuals and civilizations—from fads to revolutions.

For positive changes to occur, individuals and groups must be part of free, open-ended societies which permit a wide range of beliefs and opinions. But even in these free societies, individuals are most often raised by their parents to believe in a certain religion and to follow a certain political persuasion. Sometimes, because of the natural rebellion against parental controls or because young people learn more about the world and the forms and processes therein, their beliefs and attitudes will change— thanks in part to their use of free will.

Opinions about religion, capital punishment, abortion, human rights and a whole host of other issues have fluctuated throughout free societies, and we note that while most people have fixed ideas about these issues, there are times when a persuasive argument can alter someone's beliefs, if people are open-minded and allow new ideas to enter into their consciousness.

When the principles of adequate wisdom are applied, there is no need for a fixed belief system, since the multiple variables of existence suggest that an open-ended philosophy of life offers a path away from rigidity and conceit.

Change and novelty

While we observe forms and processes that are virtually unchanging, like the atom or gravity, there are indeed changes and novel creations along the evolutionary road. Even steadfast forms like DNA will undergo slight changes because of mutations and errors, and because of these slight changes there comes an enormous number of different species. These chance events beget novel forms.

The role of determinism is to ensure a steady repetition of forms and processes or events. Mathematical laws prescribe exact results when atoms and electromagnetic phenomena interact, with the exception of quantum effects, which cannot always be precisely measured.

The role of human design, free will, chance and synergy is to promote changes and novel forms and processes. At the beginning of biological evolution, small organic chemicals may have experienced external influences like lightening or gasses and water within volcanoes, which helped to mold these chemicals into living cells. Time and mutations created an entire assemblage of organelles into more advanced cells until we reach our current epoch, where thinking beings dominate the planet. Both chance and determinism operate to continue the species while allowing for an incredible diversity of life forms.

Free will, chance, human design and synergy allow for the wide range of novelty in music, literature, architecture, art and the wide range of changes in ideas and laws. Chance and synergy contribute to both cosmic and biological evolution of new forms.

Fanatical worship and devotion

There are many dangers arising from the worship of and devotion to a deity or idea. As if in a hypnotic state, those who fanatically worship a god or fixed idea become mesmerized by their actions and thoughts, yielding their free will to unsubstantiated beliefs. Frenzy often accompanies worship and devotional practices, a kind of mass hysteria that leaves no room for rational thought. We are often told that faith transcends rationality, but this argument is solely based upon determinism; that is, the teaching of worship and devotion passed down from hundreds and thousands of years of tradition. Followers of such tradition find their minds filled with delusions about the nature of existence. They will not examine the substance of the particular belief system; they merely follow without questioning the consequences of their actions.

Fundamentalism and intransigence are among the worst qualities of humankind, leading to a divisive and hateful world community where compassion and compromise are never practiced. The rational nature within some of us is completely subsumed by the irrational and emotional aspects of character, thus creating tensions, hatred and ignorance.

There are other kinds of worship, such as hero worship, which also can cloud the mind of the "worshiper" who can find no wrong with the hero and will follow that person or group without reservation. Devotion to these heroes must be tempered.

However, there are positive types of devotion and passion, such as devotion to one's friends, families, jobs, creative endeavors and charitable causes. One would assume that excessive devotion in these cases is not harmful, unless such devotion creates actions harmful to those outside the circle of devotion.

Devotion and super patriotism

Other dangerous forms of devotion include blind loyalty to one's profession or country which promotes the "circling of the wagons" or protecting at all costs members of one's profession, institution or country from constructive criticism.

Jingoism, the extreme patriotism that includes a fear and hatred of foreigners, is a very dangerous form of blind devotion.

These so-called patriots deny the greater good of synergistic wholes. Rather than admit to problems or errors attributable to their profession or country, the super patriots or super wagon circlers will lie or deceive themselves in order to protect profession, pride or nation.

This is an outgrowth of the need for self protection. Rather than admit to faults within, a person, a group of people, an institution or profession or a city, state or nation will create a kind of force field of lies and prevarications that seek to avoid or deflect external criticism.

In the case of super patriotism, zealots exhibit severe ethnocentrism, belligerence and a willingness to start wars with foreign entities.

If one realizes that the world community is a larger whole than each individual nation (synergy at work again), then a more balanced approach to patriotism can be achieved.

In any institution or group, it takes a remarkable individual to stand up and admit to errors, flaws, mistakes and serious blunders that could have adverse effects upon the population. The tendency of the corporation or nation is to build a wall around its organization for self protection. The exceptional person becomes a "whistle blower" and speaks out against ongoing or potential harm to the public. This action is commendable and ensures harmful practices come to an end.

Subtexts and hidden agendas

Free will and human design are certainly not always straight forward in planning events and pursuing goals. The mind can be quite shrewd in its machinations, using innuendos and subtexts to manipulate other people. Hidden agendas are not only found in politics and international relations, but in the lives of everyday citizens in business, creative ventures and personal relationships.

Many times, on the personal level, the hidden agendas are fostered by strong sexual drives that mask themselves as something else but always with the actual goal of sexual conquest. To gain favor with another human being, all kinds of subterfuges are utilized, whether based upon sexual desire or otherwise.

The personality traits and character of some people often prevent them from expressing their intentions forthrightly. It is not in their makeup to come right to the point, so some may beat around the bush and try to reach their goals by creating subtexts or just lying to themselves and others.

Hidden agendas are a part of human relationships and are not likely ever to disappear. They sometimes reflect the greed or deviousness of human nature and can adversely affect humans who become entangled in a web of deceit or misdirection.

What is luck?

All of us know what is meant by luck; we know people who seem lucky in gambling, love, and career achievements or fortunate in many activities, with an easy path through life. We say they are fortunate or lucky. (Those whose fate is darkened by many unfortunate events are said to be unlucky.)

If we believe that luck exists, then we might attribute it to circumstance or destiny, or something we cannot understand but yet bestows fortune on some people.

Those who do not believe in luck say that each person makes his or her own luck or fortune; but yet there is some nagging indication that more is at play here. Daily observances of the human condition point to a series of influences (some call it fate) that lead a person in one direction or another. It appears as though a combination of many different forces (both deterministic and chance-like) is responsible for a person leading a fortunate or unfortunate lifetime. Nevertheless, the use of free will and good public policy may be strong enough to re-direct someone's misfortune and turn things around for the good.

Winners and losers

Because of chance and determinism, all life forms experience a range of wins and losses. Winners in the genetic lottery include those whose organ systems perform with little or no dysfunction. They are also fortunate not to have any serious external blemishes or irregularities and therefore are able to maintain an aesthetically appealing countenance while having been given appealing personality traits.

Unfortunate people have inherited a range of irreversible diseases or inclinations, such as obesity, alcoholism and mental retardation. They may be raised by ignorant and/or prejudiced parents, surrounded by ignorant peers and subject to the harsh realities of impoverishment. Although it sounds pejorative to call people losers, this term is meant to show the contrast between the fortunate and unfortunate people in the world. A few such "losers" can become winners by transcending their initial conditions and rise above the muck to achieve success. There are instances where losers become winners and may become better people because of the agonies they have suffered.

Winners also include those born into comfortable circumstances, with parents, family and community members supporting one another and providing wise guidance and funds for one's journey through life.

Winners comprise a wide assortment of talented and bright individuals whose abilities are partially due to inheritance, chance, as well as free will which allows them to maximize the gifts given to them. Some talented and intelligent people may have a tendency to become smug and to flout their gifts and form cliques that snub the rest of human society.

Winners should always behave with humility, as chance might have easily made them losers in the lottery of life.

Unhappy people

During the course of anyone's lifetime, there are periods of unhappiness; moments, days, weeks or even months and years, when life appears hopeless and bleak. We know from the multitude of processes and forms that there are a great number of factors conspiring to create unhappiness, such as chemical imbalances in the brain; unfortunate physical deformity; sexual inadequacies; bullying and derision; the tragic loss of a loved one or friend; the disappointment from not obtaining approval from others; or not making enough money to purchase things that others can readily obtain.

Brain chemistry dysfunctions, including the improper firings of nerve cells as well as too little or too many neurotransmitters, also lead to depression and feelings of unhappiness. Pharmaceuticals may moderate the unhappiness, although serious side effects may dampen the relief.

The role of a society based upon adequate wisdom should recognize these unhappy people and seek them out and offer them counseling and other services to mitigate their pain and suffering.

We know what kind of cruelty can be inflicted upon people because of chance and determinism and the selfish free will of others. A truly compassionate society must do all it can to alleviate the unhappiness of others.

Absolute and relative values

Values or belief systems are most often inculcated into the young people of our population, as well as into the more mature members. This inculcation consists of passing along parental beliefs, the mores and laws of each particular society and the values held by peers and associates. These beliefs, when combined into a mental whole, form the basis of morality and conscience.

Historically and in the present day, there existed and now exist a myriad number of moral edicts, such as restrictions on what to wear, what to eat, how to pray or how to behave with other people. Anyone living in any time in history would have been subjected to the moral demands of that time and place.

The point is that even though absolute values are held at any particular time, they are actually relative values when viewed from a world perspective. Society A promotes Value A, while Society B promotes Value B. Two different value systems could be at odds with one another or co-exist with each other.

Each person holding an absolute belief will not often tolerate another belief system. But which belief system is good and which one is bad? We know that beliefs are contingent upon the laws and mores of a particular time and place. They are also contingent upon the beliefs within each family unit, peer group, each community and each institution. These comprise relative values.

But values should always be based upon adequate wisdom of the forms, processes and ideas that affect each one of us.

What are absolute beliefs?

We should all agree that beliefs and values based upon the protection and promotion of our species are good for us. Thus, rules, policies and regulations aimed at helping others succeed are absolute goals or values.

When conflicts exist between two people or among many people or institutions, the application of absolute values becomes problematic. Arbitration or judicial review may be required to ascertain what is best for each concerned party. So while protection and promotion of our species are absolute values, there are innumerable instances when judgments must be made with much deliberation. What's good for one person may be bad for another; thus we have a wide array of laws and regulations which are aimed at solving disputes as equitably as possible.

As we are contingent upon the survival of other species, many of which provide nourishment for us, a belief system that cares for all living things would appear to stand out as an absolute value. We ought to foster humane treatment of all living things, including a painless death for those creatures that provide food for us.

As we are dependent upon an atmosphere to breathe and protect us from harmful cosmic rays, a belief system that encourages the protection of the entire environment stands out as another absolute value. Clean air and water are vital to our health. The destruction of rain forests and other intrusions that remove flora and fauna for the sake of building new towns and cities will eventually come back to harm us. If humans do not practice reasonable birth control, the growth in population may overrun the planet and leave us with many more billions of starving people.(As for the future number of our species, it is now estimated that by the year 2100, there will be 10-12 billion people inhabiting the planet.)

As viewed from the whole of existence, each of us should practice pleasure and responsibility and strength and compassion.

Pleasure in this sense signifies the joys gained from the arts, crafts, conversation, play and sexuality. Responsibility means that we establish some form of order that prevents us from infringing upon the rights of others, while exercising discipline to accomplish our tasks. Strength requires us to defend ourselves and our loved ones from the occasional barbarity and evil stemming from an unfortunate combination of chance, determinism and free will. And the greatest task to achieve for any person is the exercise of compassion—to see why things are the way they are and to seek improvements for all creatures less fortunate than we. Compassion and love are a direct result of encompassing an adequate wisdom of existence.

An absolute goal would be the elimination of falsehoods and misinformation. This can be accomplished by an enlightened society which educates its members and promulgates true statements based upon adequate wisdom. An ignorant population is much more likely to wreck havoc and cause discomfort and pain than one that is well educated. A campaign to spread the truth is based upon early education that presents the forms, processes and ideas and shows how each person or group is contingent upon the other.

It seems, at least now, we are not going to witness such an internationalization of truth.

Regulations & social engineering

All social structures contain rules and regulations. Whether it's a family, organization, corporation or nation—there are always codes to follow to maintain and sustain the structure. In sub-institutional groups (like friends or family), regulations are often lax and permissive, allowing for a wide range of behavior as long as that behavior does not jeopardize the members of the group. In large, bureaucratic institutions, like corporations, the government and the military, the regulations are stricter; those who do not follow rules are often punished or disciplined, demoted or dismissed.

In the varied types of social contracts, the individual yields some of his or her freedoms and actions to receive the benefits that the structure offers.

Nations, with countless institutions, require citizens and groups to follow laws, codes, edicts and regulations that have accumulated over the years.

In a nation ruled by a dictator, the rules and regulations primarily benefit the ruler or ruling group of despots. Self-serving social engineering is quite prominent in these tyrannical states.

In a democratic nation, there are reasonable regulations that allow citizens to participate in a wide variety of activities and even voice different opinions with impunity.

In socialistic leaning nations, there is greater social engineering which promotes the design of many programs that provide health, counseling services and other opportunities for the betterment of all people.

In capitalistic leaning nations, there is some significant resistance to social engineering and a demand for free markets and the profit motive to determine what is good and bad for the country. Rugged individualism fights against the design of social programs, claiming that each person must pull himself up by his bootstraps without assistance from others. There exists the belief that social engineering is an attack upon their freedom. This is a narrow and, frankly, ignorant stance to take because of the synergistic interplay among design, chance, determinism and free will. Unfortunate people often need a hand up from the mess in which circumstances have placed them.

As we know from the effects of bad circumstances, there will exist those individuals who commit anti-social acts, like fraud, tax evasion, hacking and online credit card theft as well as outright vicious crimes against people. The role of government in these cases is obvious—to protect its citizens by removing those who wish to harm the whole.

But there are always the proverbial questions—more or less government, more or less regulation? What's more important—the individual or the state?

In looking at forms and processes, we know that the government is the synergistic combination of all interdependent people and groups. If the government follows the path of adequate wisdom, it will issue policies which promote the general happiness of its citizens while effecting rules and regulations that protect its people, groups and institutions. Social engineering projects include interstate highways and traffic rules; airline regulations; social welfare programs; universal health care; food, drug and environmental safety programs (including clean air and water); education grants; and scores of other projects that improve the welfare of its citizens.

It is a fine line between individual rights and government regulation, an ongoing interplay of forms and processes. Institutional designs and deterministic forces do battle with the free will of each individual. In an ideal situation, both forces act to benefit each person, as well as the group, institution and state.

In the current civilization, nations have grouped together to form communities with shared currencies and regulations and courts. A prominent example is the European Union. In addition, the United Nations ostensibly represents the global community and all people of the Earth. The UN has been hamstrung by some political groups and nations which refuse to yield some of their sovereignty to a greater entity. As we have discussed, the synergistic trend in evolution suggests that eventually a world government of some sort will emerge.

In the meantime, governments everywhere ought to guide themselves according to adequate wisdom, by providing strength and compassion, as well as pleasure and responsibility for all people.

The most important factor in governing is to promote the health and welfare of citizens, including fair treatments of all people; to encourage educational development; to encourage the participation in pleasurable activities (including the arts); to provide for an infrastructure of safe roads, buildings and bridges, among many other such actions.

Finally, if government understands that chance and determinism play an enormous role in the lives of citizens and the myriad events in their lifetimes, then compassion for the unfortunate people whose lives are wrecked by circumstances beyond their control must become an overriding concern.

Sovereignty of nations and a world government

All civilizations have exhibited a patchwork assemblage of nations, each following a unique ongoing system of laws and mores. Each nation stands as an individual whole, competing with, trading with, or at war with each other. Throughout history, stronger nations have subsumed other nations and have tried with various degrees of success to establish empires. In the current civilization, once again we see a patchwork collection of nations but with some major alliances and confederations, such as nations tied together in economic, legal or military alignments. In these synergistic arrangements, some sovereignty of each nation is yielded for the betterment of the member nations.

However, in order to provide a universal command center that promotes the values of adequate wisdom for all people, each nation would have to yield more of its economic, political and military strengths. This concept of a world government is found in the United Nations which tries with varying degrees of success to enforce its own regulations among the nations of the Earth, while also accomplishing many good deeds through its humanitarian agencies.

We encounter the tension between individualism and synergy, so at this point along evolution, few nations will substantially yield to a central nucleus, so to speak, that could establish universal goals to protect the flora and fauna, the air and the water of the planet and the downtrodden. Could there be a time when all nations coalesce and work in cooperation with one another to promote the flourishment of the species? In this day and age, it seems highly unlikely that such a phenomenon could occur. But we cannot rule out the eventuality of a synergistic civilization that works on behalf of all people regardless of economic and social status.

What political and economic systems benefit humanity?

Political beliefs and economic systems dominate the time and place that individuals find themselves. Historically, we have seen any number of such institutions that structure the society. In today's western European democracies we find a mix of socialism and capitalism, with more accent on socialism than capitalism. The reverse is true in American society. In addition, there exist nations without any forms of democracy, ruled by benevolent or authoritarian dictators (usually the latter).

Purely socialistic forms of government may hamper individualism. Purely capitalistic forms of government tend to make citizens subservient to the power of corporations and profit motives.

Common sense would dictate that political and economic systems ought to protect and support its citizens, its flora and fauna and its atmosphere and water, while not harming other people in their nation or in other nations.

Political and economic systems must also allow for individual freedom of choice and action, as long as these choices and actions do no harm to others.

Political systems must encourage scientific and technological research as well as the development of the arts and humanities.

There remain sharp schisms among political groups and alignments, some whose sole aim is the debasement of government and the eventual end of a paternalistic society where states and nations take care of the sick and the poor. These extremely conservative political groups contain ego-based rugged individuals who expect our population to function as survival of the fittest, an old concept that has long been discarded and which runs in the complete opposite direction of adequate wisdom and compassion.

What is right and what is wrong?

We regard as right or good those events or things which promote successful actions or feelings, which lead to the flourishment of the species, which make us feel happy and which tend to help others achieve happiness. Doing right or good is the hallmark of compassion and love, the goal to which all people ought to strive. Doing right comes from the balancing of pleasure and responsibility and strength and compassion.

Goodness transcends political and religious creeds. Totalitarian regimes and dogmatic religious prescriptions cannot supersede the essence of doing right. Historically, morality has varied from civilizations to nations, to laws that govern people. Some laws are not right when they arbitrarily impinge upon individual liberties.

We can cleverly say that what is wrong is not right. So virtually anything that inhibits or represses or disables, diminishes or condemns humans or human activities or the environment in which they live can be considered wrong.

There is also the idea of amorality, which suggests that there is neither a right nor wrong action. An earthquake (while bad for those harmed by it) is not an event about which one can assign any form of morality. If one believes in complete determinism, then all actions—no matter how good or bad they appear to be—are amoral and cannot be judged nor condemned. Under this system of belief, murder, rape and theft would be the result of chance and determinism and not judged as immoral, assuming there exists neither free will nor human design.

Actions arising from the free will of the mind can be classified as bad or good based upon the tenets of adequate wisdom. However, when actions are the result of the totality of free will, chance, determinism and synergy, then a holistic morality ought to apply which moderates the polarity between good and bad.

What is evil?

In addition to being wrong, evil is regarded as acting out of malice and performing acts of violent or heinous crimes against living things. Is evil inherent in each of us? At least one major religion believes it is. But it appears as if evil is generally a combination of defects in the brain chemistry and a series of unfortunate entanglements in the life of an "evil" person.

Evil and/or bad actions occur in every historical period, in every civilization, and in some institutions or families and individuals. They recur and are most likely perpetuated by other recurring forms and processes. It would be naïve to believe that we can rid the world of badness or evil as these are recurring features of humanity (the manifestations of chance and circumstance) and will persist throughout the ages. In this sense, evil is the manifestation of an unfortunate confluence of certain recurring forms and processes.

In today's world there are fanatical organizations that brainwash many of its members to commit unspeakable acts of terror. Irrationality combines with design and determinism to produce people who strap themselves with explosives to kill others and themselves because they have been led to believe that their actions will further the cause of their group.

All of humankind is not inherently evil; only a small percentage of humans are affected by the deterministic or random processes that give rise to the circumstances that create the occurrence of evil actions. Is there some evil or badness in all of us? Probably yes—in varying proportions, such as an occasional evil or bad thought or action. But for the most part, social policies and a compassionate conscience keep us in tow.

Why do we like to blame others?

Humans like to assign blame for errors, mistakes or bad actions. This assignment of blame forms a feedback that in some cases helps to prevent errors or bad actions from recurring. But assigning blame places us on a very slippery slope, since many actions may result from determinism and chance, and thus the person may not always be to blame. A good rule of thumb, before assigning blame, is to think long and hard about other factors that may be at play. Instead of criticizing people for all their assumed mistakes, it is a far more humane effort to try to understand them. Human beings are fragile; life is difficult for many people. Assigning blame may yield a result that is not helpful at all.

The "blame game" is common among the population. It is often used to deflect one's own errors or mistakes onto to unsuspecting, innocent humans who are unaware of their accusers.

A great many people refuse to accept responsibility for their actions and look for any excuse to exonerate themselves. However, we do know that many of their actions are the result of chance and circumstance, so once again we are stuck between events generated by free will and those influenced by determinism. This is one of the greatest human dilemmas, and it seems better to err on the side which moderates our anger against people whose actions ostensibly appear to be the result of free will when in actuality may germinate from amoral determinants.

Why do we die?

All living things die. Some die by accident, some from disease but most from old age. We die when the heart and brain completely stop functioning. The older we become the more likely it is for our cells to break down. Depending upon what laws prevail, some people may die a terrible death, without the recourse to euthanasia. Others may die peacefully in their sleep. Just how we die is the result of the interplay between chance and determinism (and free will as well, should we be permitted to control the end of our lives.) There should be no debate about euthanasia. Neither government nor religion should ever interfere with the free will of a person suffering from a terminal illness who wishes to end his or her life.

Our cells act as time clocks, with tiny bits of DNA, called telomeres, which cap the ends of DNA strands in the chromosomes, preventing them from fraying and dying. But, as we age these telomeres shorten and cells begin to die.

At some point in the lives of all humans, the fact that we will die strikes itself upon us like a lightning bolt. Those who take comfort in religious explanation feel satisfied that death is not the end of life but merely a transition step to another form of existence, whether rebirth in another form or a soul "living" in another dimension.

Others in our population hold that death is the final cessation of life and no other forms we could recognize will occur. They believe that death is the final stop. They say that belief in an eternal soul is a pie-in-the-sky balm to soothe the anxieties of all of us facing impending death.

Of course there is no proof for either assertion, so this argument will doubtless recur for as long as human beings exist.

One certainty is that upon death, consciousness ends and the life form and its cells and organ systems no longer function.

Why is death part of evolution? What benefits occur that promote the death of individual living things?

During the course of the evolution of life forms, we see that nature reshuffles the deck by creating a specific lifetime range for all species. Some botanic forms live thousands of years while some insects live a day or two. Human lifetimes have been increasing throughout the centuries thanks to innovations in medicine, nutrition and technology. An eighty-year-plus lifespan is no longer thought to be rare. The human lifespan will doubtlessly continue to expand.

There are some scientists specializing in gerontology who believe that the human being is capable of living much longer. A few of these scientists believe that it may someday be possible for humans to live for hundreds of years, assuming certain fixes in cells can be accomplished, thus allowing cells to replicate as cleanly as they did when the individual was in his or her youth. A very few of these scientists think that immortality might be possible in the far future, assuming we can conquer cancer. The longer we live, the more likely cells in our body will mutate into cancer cells and eventually do us in. A cancer cure could signal the advent of very long-lived humans.

But at this moment in the history of our species, we are stuck with a finite lifetime. We know how we die, but why do we die? Death has come to all living things during the past 3.8 billion years. Species die and new ones are born. If death did not occur, eventually there would be no more space on Earth to house all living creatures. So death accomplishes this reshuffling of the deck and allows newborns to occupy available space. Death thus provides room for new life forms to evolve (which is of no consolation to the living!). Death also provides fodder for future generations, acting initially as fertilizer and eventually as a possible energy source.

Acceptance of the eventuality of one's death is not an easy task. Even with the balm provided by religions, everyone understands that death terminates the palpable sensations of love, joy, sexuality, humor, taste and touch; as well as the mental stimulations provided by conversation, thought, literature, art, films, music and all the wonders of life.

But nature assists in the acceptance of death by making us old, weary and sickly, so that many of us may look forward to an end of pain and suffering. Some unfortunate people, at any age, take their own lives because of a series of chance and deterministic events inside the brain and within the external world.

There is one final thought about the extension of life and the problem of space available on the planet. While science fiction now, there is certainly the far future possibility of life extension and the terra forming of planets, moons and other celestial bodies, someday allowing populations of humans (and animals and plants) to live off-planet. Terra forming is the creation of or changing the atmosphere of a planet or celestial body to make it hospitable to life forms. It is a scientific possibility but would take an inordinate amount of time for the process to become effective. Also, the far-future notion of humans building space colonies cannot be dismissed.

As for the near-future health of humankind, much is being made of the harm we are presently doing to the atmosphere of Earth and the strong possibility that if we continue to emit excessive amounts of pollutants into the air, we face the specter of significant climate change to a point where human life is no longer sustainable. Nothing exemplifies how infantile the human race still is than its reluctance to deal with the longterm preservation of its species. Infants want immediate gratification; so do many present day political leaders around the globe who cannot see a foot in front of them! It does not take much wisdom to see that we are teetering upon a dangerous point in our evolution. If immediate steps are not taken, there remains the strong possibility that we could destroy much of the life on our planet.

Assuming we wake up and initiate the necessary actions to reverse or stabilize climate change, there remains one undeniable fact—so far off in the future that none of us needs to concern ourselves—and that is the eventual end of planet Earth.

In a few billion years, the Sun will expand and grow hotter, eventually changing into a red giant star, vaporizing the inner planets, including Earth. Even if far-future humans leave the planet and colonize farther away, the Sun eventually becomes a white dwarf star losing all its heat, and thus freezing all the planets remaining in the solar system.

There are possibilities that would enable very far future humans to survive. One is the creation of massive fusion energy heat sources that replace the heat from the dying Sun. Also, in the very far future, who knows what wonders genetic engineering can accomplish, such as the astounding notion of the creation of new human life forms, with an advanced central nervous system or advanced organs and cells that can live in extremely hostile environments and perhaps even exist as autonomous beings without completely depending upon external energy sources!

Also on the far future horizon, the Milky Way Galaxy is due for a collision with its dwarf galactic neighbors, the Magellanic Clouds, in a few billion years, and a collision with its large neighbor galaxy, the Andromeda, in more than five billions years.

So we know the very, very longterm future of life, and it is not a rosy picture! Fortunately, we have an incredible amount of time remaining, assuming that we do not ruin our atmosphere, run out of food for the expanding global population or experience a catastrophic collision with an asteroid or suffer from any other astronomical, biological or geological disaster.

Death: The Ultimate Challenge

No matter how much we enjoy life and suck the marrow out of the joys we experience and the accomplishments we achieve, there comes a time when we must reconcile life and death. As overwhelming an experience as life is, so too is the challenge of our own mortality. How do we face death?

The loss of a loved one (even a pet) can be one of life's most devastating crises. Years and years of companionship and common ties are suddenly ended, and the surviving individuals find it very difficult to reconcile their loss and regain some semblance of human activity during and, in some cases, after the grieving process.

The rational mind remains befuddled by the realization that each one of us will someday no longer be part of the universe. Such a statement is painfully obvious and yet poses our greatest challenge somehow to accept our own finality.

There is no denying that belief in religion has helped an untold number of humans face death throughout history. Whether religious beliefs are founded or unfounded, they do offer solace to those who believe in a soul or afterlife. The truth about death is not an apparent factor in belief. That is, if it is true that there is neither afterlife nor soul, it matters not to those with faith. The act of believing soothes the anxieties associated with the death of a loved one or the impending death of any individual.

If we subscribe to an adequate wisdom of existence and therefore do not necessarily subscribe to a belief in a soul or afterlife, how do we handle or accept the finality facing each one of us? Such acceptance requires remarkable strength and wisdom to grasp the notion that we will no longer be part of life.

We understand that our species will continue to exist and that any contributions we made during our lifetimes will continue to hold sway among the population for some amount of time after we are gone. If we have lived a purposeful life, one filled with good deeds for our families and species, then we know we have made a mark on existence. We continue to live, as it were, through our accomplishments and through our families and through our children and their children ad infinitum. In the age of the internet, we all can sustain an eternal digital lifetime through social networks that contain our biographies, our photographs, our videos and all manner of information about ourselves and our ideas and accomplishments when we lived.

But whatever solace one chooses to deflect the imminence of death, deep down in our consciousness we know that the universe will continue to exist while we do not.

Those who believe in a hidden meaning to life and death will counter with arguments usually calling for a continuation of each life force after death. They often invoke a form of the physical law that matter cannot be destroyed, but rather transformed into another substance. This is all well and good, except for the fact that all living tissue in our body, from neurons to blood cells, no longer function and begin to decay and die. It is difficult to imagine a soul devoid of memory and intelligence.

But, as we have seen through these many essays, not all is certain about existence. Could such a life force exist? Could such an unknown force carry our memories unto another level of existence about which we know nothing?

The adequate wisdom answer is a cautious no. But we can never be certain, and if such a belief in the hereafter provides succor for those who lose a loved one or face death, then far be it for us to dissuade them.

Each one of us reconciles the loss of a loved one or our own appointment with death in our own manner. It is the ultimate challenge.

So how do we understand life and existence?

We think that there is a unique universe or that there is an infinite number of universes. We think that either existence has been designed by some unknowable entity or has been assembled randomly to constitute just the right amount of fine tunings to create life. In either case, forms and processes are determined to unfold along with chance events that create varieties of structures which have a tendency to form greater wholes.

Whatever forces and forms coalesced to produce living things, we know that the physical universe existed billions of years before the advent of life forms on Earth. And several billion years existed before human beings evolved into their current state. So here we are today, with the gift of cognition allowing us to speculate about the nature of existence—from the ultra small to the ultra large.

To understand the world, we have created tools by which to make observations about structures and events, whose observations turn into ideas. We sort these forms, processes and ideas into physical, biological, human and social categories and suggest that all such categories be included in any decisions and judgments.

We recognize that free will and human design do not exist unimpeded, as chance and deterministic events also direct our policies and judgments. We understand that the trend toward synergistic combination of forms and processes suggests that all facets of existence must be grasped before we reach an adequate wisdom of the world.

We then strive for pleasure and responsibility, as well as strength and compassion. We seek flourishment for our species, all living things, as well as the atmosphere, geosphere and hydrosphere from which we derive sustenance.

Is there meaning to existence?

Is there purpose or meaning to the universe, to galaxies, stars, solar systems and living things?

We certainly know that there is a WOW! factor when we think about the forms and processes that simply amaze us—whether they are the intricacies of protein formation in living cells or the strange black hole in the center of our galaxy; whether they are the color and smell of an orchid; the beauty of a snow-capped mountain; or the enjoyment obtained from physical and social intercourse. We are (or should be) in a constant state of amazement that we exist and are part of the enormity of all these things.

But is there any meaning to any of this?

If we believe in cosmic design or preformation, that is, a prior blueprint for all that exists, then we could posit that there is a goal or purpose to everything. We could call this design or preformation an act of a God which directs the evolution of particles and atoms into stars and planets and living beings. Whether such a deity exists, it is our minds that assign a purpose or meaning to existence. We develop various creeds based upon the human interpretation of God. We have faith that somehow all that exists has purpose, has meaning. A majority of the human population believes that what occurs is meant to be, that even though God is an ineffable and mysterious concept, we exist as part of a plan or a purpose.

Or, we could dispense with a God and yet still believe in preformation as a result of built-in mathematical laws peculiar to our specific universe which directs the forms and processes. In this case, there is no discernable cosmic purpose or meaning.

We might believe only in chance and determinism which would negate meaning as we understand it. The universe might have evolved randomly and thus has no built-in design and therefore there exists no cosmic meaning.

As humans, we may have a feeling that a void inside of us exists, which calls for a reason why things are the way they are. Everyday activities are geared toward a goal, toward some outcome that does indeed provide meaning. Therefore, each human lifetime is goal oriented, and so while there may not be any discernible cosmic meaning, there is plenty of meaning in our lives. We live to create, to love, to work, to enjoy, to reproduce, to do a great many things, and thus meaning can exist and flourish.

In regards to free will and synergy, our minds can develop an adequate wisdom that helps to explain the holistic nature of the world. We know we cannot understand why cosmic and microscopic forms and forces exist, but we cannot say with certainty that a galaxy, for example, has no meaning. But if there is such a meaning, it is well beyond our comprehension.

And if there is an infinite number of universes, then anything and everything is possible. Admittedly, it is quite difficult for our minds to grasp the notion of infinity, which calls for an endless spacetime; but even more difficult to grasp is the idea that the universe actually began from nothing. The more likely scenario which is backed up by quantum mechanics suggests such an endless world exists and has and will always exist. The implications of eternity tend to rule out the idea of a creator, but we cannot say for certain that this is the case. As stated often in these essays, there could be forms and processes of which we have no comprehension. It's difficult for us to admit that we do not understand the cosmos. Our pride is at stake when forced to admit that we have doubts.

Finally, many religious leaders claim that if people do not follow the worship of God in a predetermined manner, then there can be no meaning in the world. This notion is an outright effrontery to common sense and adequate wisdom. A non-religious person can have as much or more of a meaningful relationship with the world than a religious person.

Meaning comes from the appreciation of life, of the universe and the good deeds that a person performs while a member of the species.

The future of existence and the Earth

On the grand scale, we have already discussed the possible fate of our universe. All the galaxies and clusters in our universe may eventually experience a heat death in hundreds of billions of years or they may eventually coalesce into a universe-sized membrane that collides with another brane and form a new Big Bang and a new universe—another synergistic form. But we have also noted that there is a strong possibility that the universe is infinite, without beginning and end, and thus existence will never end.

As for the Milky Way Galaxy, it faces a collision with the Magellanic Cloud galaxies in a few billion years and another collision with the Andromeda Galaxy in about five billion years. In both cases, the distance between stars being so great that it is quite possible for only minor disruptions to occur from these collisions. The upshot of these events is the creation of a new synergistic whole, a brand new galaxy, Milky Way Two (or Andromeda Two), one of perhaps 150 billion others.

As for the Sun, which is about five billion years old, we believe there are another five billion years left before it burns out its nuclear fuel, but over the next couple billion years the Sun will become hotter and larger with serious consequences for Earth and the solar system. Then later, the Sun will eventually burn out completely, providing no heat source for the solar system.

As for the Earth, there are many scenarios in the short and longterm. The build-up of carbon dioxide in the atmosphere could seriously damage the planet. In the worst case, a runaway greenhouse effect could make life on the planet impossible except for some small organisms that can flourish in high temperatures. Just how long it will take conditions to kill off animal and plant life if carbon emissions are not drastically reduced is difficult to assess.

But it is possible that less than a few hundred years from now, unless we take dramatic steps, the human race could perish or face bizarre and unbearable natural phenomena.

Although still in the realm of science fiction, it might be advisable to start thinking about planning a serious effort to terra form other bodies in the solar system and to prepare space stations that contain the seeds of life for use on man-made or extraterrestrial bodies.

There are other threats to Earth, such as collisions with major astronomical bodies large enough to cause great or catastrophic damage to the planet. And there is always the threat of nuclear holocaust, although trends lately have been to reduce this threat significantly. And there is always the danger of some airborne virus doing tremendous harm to living creatures, including humans.

Then longterm, we know that the Earth will not always exist because of the Sun's irrevocable enlargement process; yet we need not fear this for a billion years or more.

In terms of the future of humanity, we can barely guess what will occur in 30 years, let alone a million. But because of evolutionary change, occurring in more or less millions and tens of millions of years, we as a species could easily become supplanted by a brand new species, branching away from Homo sapiens, a new, advanced species, possibly called *Homo provectus*. (An advanced form of Homo sapiens.)

Creating a life plan in the context of forms and processes

A partial life plan is already mapped out for each person. Being born with a specific genetic package with predispositions built in presents us with a physical and mental form that has been somewhat predetermined by our DNA and that of our species. That is, our appearance and many body functions have been planned for us. Certain assets and vulnerabilities have been given to us through heredity.

Whether we are fortunate or unfortunate in the genetic lottery is mostly out of our control. Certainly some setbacks, like poor vision, weak teeth, mild skin problems and other manageable physical drawbacks can be adjusted and often eliminated by modern medicine and technology. Even more serious physical and mental setbacks are now being addressed and mitigated by medical science. Nonetheless, the body and mind, including creative tendencies and inclinations, are passed along to us; this genetic determinism must be part of our understanding.

In addition to the genetic lottery, we are subjected to the circumstances into which we are born and raised. Creating a life plan is greatly influenced by the family to which we belong; the time period and location of birth; peer groups; political and social arenas in which we find ourselves; the financial and religious background of our families and communities; and a wide variety of occurrences, happenstances and crisscrossing events.

But despite all the factors that mold and direct us, there is always the mind that can reflect and design a better lifetime by creating benevolent social policies and individual ideas that transcend some of the negative determinism and chance events intruding into our lives.

The task of trying to understand life and the universe can overwhelm virtually all of humankind, but attempts must be made, for the more we understand, the more we can reasonably establish our life plans which resonate with the workings of the world around us.

It must be stated clearly that living a life based upon "gut" reactions can work out just fine for some people who use common sense and are fortunate enough to be free from debilitating drawbacks. Like driving an automobile without understanding how it works, people can move through life automatically, so to speak, using their wits and common sense to get by.

But an attempt to find a deeper understanding of the world around us brings us closer to the creation of an even more effective and rewarding way to live our lives. And how do we seek understanding? By asking and answering questions! By looking at the forms and processes and ideas swirling around us and realizing that we are all part of everything else.

A SENSIBLE LIFE PLAN

A sensible life plan allows each person to base his or her decisions upon the multi-faceted array of wholes (structures) and processes (events) that take into account design, determinism, chance, free will and synergy. A life plan that recognizes the workings of the universe, including human affairs, allows each person to act with strength and compassion, pleasure and responsibility.

Strength is necessary to protect oneself and one's friends, family, and world community. It also allows us to act with tenacity in all our endeavors.

Compassion requires sensitivity to all people and all living things. Compassion results from understanding the trials and tribulations of other people. Compassion is based upon the adequate wisdom we have accumulated.

Pleasure allows us to enjoy the bounty of nature, of sexuality, of humor, joy, arts, conversation, and the rest of life's treasures.

Responsibility requires us to establish guidelines that keep us in check with the world around us. Responsible people organize and categorize to find meaning in the world. They balance their lives so that most actions are moderate and belief systems are open to new ideas.

A productive life plan allows each member of our population to seek clarity about the world and perform good deeds for oneself and others; to engage in thoughtful debate about the nature of existence and its forms, processes and ideas; to enjoy the company of one's friends and family and associates; to temper one's emotional judgments of others; and to rise above any negative influences of chance and determinism.

PART EIGHT:

FINAL THOUGHTS

This last part tries to summarize the key points made in this work, as well as present various ideas, questions and speculations that lead us toward the formation of an adequate (workable) wisdom.

Hopes and aspirations

Having looked at and examined an overview of the formidable universe, including the lives of living things, we are humbled by the extreme profundity of existence. Many different conclusions can be reached, depending upon our points of view, yet in the final analysis the rational mind realizes that many matters remain unsettled and that only an adequate wisdom is achievable.

Whether we believe in a God, whether we believe in chance and circumstance or whether we sustain doubt about all of existence, we can still entertain the notion of hope—hope for a fulfilling lifetime, hope for good things to happen, and hope for progress to improve the lot of many downtrodden individuals.

To hope is human. It springs forth from our optimistic nature and asks for meaning and accomplishment in each person's life. Not to hope is also human and may produce a doom and gloom scenario for humanity. In both cases, personality traits and life experiences greatly affect one's propensity to find the glass half-full or half-empty. But for those who aspire to accomplish their goals in life and to achieve any kind of success and prosperity, the need for hope remains strong within them.

Without hope, existence can be cold, empty and meaningless. With hope and aspirations for a better world, humans can move through life as gracefully as possible.

Recap and review

The brief essays in this work stressed that forms, processes and ideas are inexorably intertwined and that no idea can make full sense if it excludes all other forms, processes and ideas. We listed the twelve variables of existence and suggested that all of them affect the universe and all living things. We noted that there exists the recurrence of physical, biological, personal and social forms and processes that continually interact with and affect all the components of existence.

We pointed out that the mind is capable of forming wise judgments and policies even though it is subject to the powerful influences of emotions, personality traits, self-interest and formidable social and political structures.

We reached the conclusion that there are a number of recurring forms and processes and ideas which continually influence human behavior and events. Because of recurrence, there is repetition of personality traits and emotions, with parts of history repeating itself, excepting the wild cards of free will, human design and synergy, all of which initiate novelty into the human experience.

We suggested that the issue of a linear evolutionary trend of existence is not completely settled, and as such the concept of preformation cannot be absolutely dismissed. Of course, the forward progress of evolution seems the most logical approach, but there appears to be more than meets the eye when it comes to explaining existence and the universe.

One main theme throughout the essays is the apparent duality between individualism and synergy, prompting the emergence of forms and processes which carry out specific tasks while at the same time existing as parts of a greater whole. Cosmic and biological evolution creates steadfast forms which nonetheless evolve through chance and synergy into greater forms.

Perhaps the universe represents the Grand Synergy of all forms and processes, and it continues to evolve. New stars, new planets and new living things are continually emerging, brought about by the fundamental processes that remain constant, as well as by chance events that create novelty in the cosmic and biological realms and by free will and design in the human world.

Also suggested is the creation of a holistic morality which tempers the extremes of right and wrong, since human actions are not only based upon free will but also upon the amoral factors of chance and determinism.

We have discussed some absolute beliefs that are based upon the protection and promotion of our species, the biosphere, the geosphere and the atmosphere. We have said that individuals and institutions should promote pleasure and responsibility, while practicing strength and compassion.

We know we must work to eliminate falsehoods and misinformation that tarnish the everyday life of people around the world. For this to occur would require managers and leaders who promote universal education for all citizens of the world. Yes, a gargantuan task to say the least—but one absolutely necessary for the intelligent and compassionate evolution of our species.

But throughout all human lifetimes, it seems impossible to suggest seriously that only goodness will prevail and that one day we can figure out how to educate and feed and shelter the entire population. And the reason we will not reach those goals in the near future rests with some recurring forms and processes that bring about the churning and antagonistic features of human personality and character. The selfishness of some of our species is well documented. We mostly run past one another in a reckless fashion; that is, we are mostly concerned with our own selves and care little for others who require our attention. So the consummation of a well managed humanity perhaps lies in the far future.

Because of the synergistic nature of existence, we find that we humans are much greater than our DNA and organ systems. In fact, all new wholes are different from their parts, and so a holistic view of the world makes much sense, while not discounting the individuality of all forms and processes.

We have encountered the strong possibility, at least according to theoretical physics and quantum mechanics, that there may very well exist an infinite number of universes, which may lead us to a never ending world, a world that is eternal; that has neither beginning nor end. If this is true, then the concept of God becomes questionable. That is, with no beginning and no end, there may not be a need for a creator or need for a prime mover or maker of all things. An eternal universe would help solve the riddle of space-time, since it will have always existed and will continue to exist forever.

Also, contemporary physics and cosmology lead us to the possibility of a non-local universe, one that may be somehow interconnected no matter how far the distances are from one another. As such, the holistic view of existence takes on greater significance in our view of the world.

As for religion, we noted that it provides comfort and meaning for its members as well as a psychological cushion on which to rest fears and doubts. But we have seen that in the case of any absolutist system of thought, there is a rigidity of moral standards that can create fanaticism. Those who advocate for religion claim that without a God there is no moral accountability and that nothing makes any sense because no one is in charge. The main problem with this line of thought is that humankind is thus relegated to a belief in a supernatural being, so instead of trying to take charge of one's life, each human is urged to follow the dictates and moral edicts of many different religions, most of which offer contradictory belief systems.

Since most religions admit to the ineffability of God, then who creates religious creeds if not mere humans, who are of course fallible?

In a synergistic world, humans can make wise choices based upon adequate wisdom and lead purposeful lives without the interdiction of religious dogma. In fairness, it must be admitted that some religious creeds help to promote compassion and goodness.

It is necessary to point out that people and groups can hold contradictory viewpoints about the universe, existence and God, as long as they are willing to concede that their ideas reside in an area devoid of concrete truths and as such only *allow for the possibility* of their belief to be accurate while not shredding competing ideas.

THE PROBLEM WITH DOGMATIC VIEWPOINTS
We have also noted that there are those who pontificate one particular point of view to the exclusion of all others.

A physicist spends a lifetime studying particles and becomes so absorbed with them that it becomes natural to believe that the world, including human behavior, is simply a collection of particles that can describe all of existence.

A biologist spends a lifetime studying the mechanisms of cellular functions and DNA transmission and becomes so absorbed in them that it seems natural to declare that life forms are the result of only chance and determinism.

A chemist spends a lifetime studying the properties of atoms and molecules and becomes so absorbed with them that it seems natural to assume that all of existence is simply the formation of more and more complex atoms.

A neurologist spends a lifetime studying the neural activities in the central nervous system and believes that human free will is an illusion.

A clergyman spends a lifetime studying and pontificating on a number of religious texts and ideas, while proclaiming that God provides the answer to all our questions and problems.

Now certainly there are many specialists mentioned above who hold wider viewpoints, but the point is that many people wear blinders when espousing ideas about existence. They fail to see the complete context, the unified whole that encompasses the world. The forms, processes and ideas reflect the physical, biological, personal and social spheres. All mesh together and cannot be separated, except for heuristic purposes (aids to learning about existence).

There are many other frames of reference that encompass the nature of existence. There are the geocentric and heliocentric views of the celestial sphere; there is string theory and the classical, quantum and relativistic frames of the physical world; there are the sociological and psychological frames of reference; there is the cosmological view of existence; and many others. The totality of these frames needs to be seen as constituting a unity. Only by grasping the entirety of existence can we begin to establish an adequate wisdom and design life plans for individuals and social policies for groups, institutions and nations.

THE MODES OF EXISTENCE
We have looked at forms and processes through several modes—design, chance, determinism, free will and synergy. They are all major factors in studying the existence of forms, processes and ideas. These modes are briefly discussed in the following paragraphs.

On the cosmic scale, design may suggest built-in instructions within mathematical and physical laws that permit particles and atoms to form stars, galaxies and life forms. This design possibility does not mean that a Maker with a purpose created the world; it simply suggests that chance may not be the only cosmic determinant, and that preformation inherent in mathematical and physical laws may exist.

In the biological world, living things can design other things, enabling them to construct forms that have a purpose—such as a beehive, an ant hill, a spider's web, or the ideas and structures created by humanity.

On the cosmic scale, chance events combine with other physical forces to create (possibly) an infinite number of universes, the diversity of stars and galaxies and the creation of planets (including plate tectonics) and other non-stellar forms.

On the biological scale, chance has a tremendous effect upon evolution and the diversity of living creatures. Chance also operates when the sperm and egg cells are fertilized during recombination of genetic materials, along with random glitches in the genetic fiber.

On the human scale, chance affects the circumstances of each person, the parents, the locality and the confluence of all randomly interconnected forms and processes.

On the social scale, chance events in history helped to mold civilizations, create geological boundaries and influence the lives of individuals and groups that formed nations.

On the cosmic scale, determinism represents the prime physical forces that construct particles, atoms, molecules and stars. All causal events are deterministic. The production of heat and light and electromagnetism from stars are deterministic events. The formation of the 92 natural elements and countless molecules are sorted out by determinism. (However, if we were to admit the possibility of a preformational nature of existence, then we could say that determinism is simply the means to an end already pre-scribed.)

On the biological scale, determinism is responsible for the faithful reproduction of cells, proteins and enzymes following the strict coding of DNA. When errors occur by chance in the DNA structure, determinism constructs cells according to the newly modified DNA.

Biotechnology and genetic engineering are powerful new design tools created by human thought and volition that allow for genetic sequencing, the creation of new life forms and the foundations of cures for many diseases.

On the human and social scale, determinism follows design, so that instructions are carried out according to plan. Architectural drawings and computer programming are designs that set in motion determinism faithfully to reproduce the plan. In addition, certain rigid social and religious laws and rituals produce a deterministic course until otherwise changed.

On the human and social scale we find free will, either emanating from individuals who write, plan, compose, design and think; or from social groups that represent a meeting of the minds, allowing for a combination of free wills to form group ideas, projects and social policies. Human design and free will are often interchangeable aspects.

On the cosmic scale, synergy encapsulates the physical forces of gravity, strong and weak forces; as well as electro-magnetism and quantum mechanics to create a series of additive forms, from quarks forming neutrons and protons (greater wholes) to electrons and the atomic nucleus forming greater and greater sized atoms. The combinations of the prime forces create stars. And then stars combine to form the much greater galaxies. There are physical hierarchies, as well as biological hierarchies. The cosmic and biological evolutionary trends have followed a synergistic route.

The animal and plant kingdoms on Earth exemplify the hierarchy created by synergy. In human biology, the stem cell is programmed to create special tissues that form organs which comprise the organism.

Not only is each human a synergistic whole from within but also from without—as part of an environment, a family, a group, an institution, a nation and the world community.

CELESTIAL INFLUENCE

To the consternation of theologians, scientists and social scientists, it has also been suggested that one cannot disregard the influence of our solar system and of celestial influence on the Earth and its physical and biological forms. After all, the solar system represents a significant synergistic whole embracing an ever changing dynamic and cannot be excluded from an examination of the forms and processes. Quantum mechanics allows for action at a distance, so the entire solar system (and the universe) represents an ever changing field influencing all its constituents. (Please refer to the essay on *What is Personality?* and the references to the interconnectedness of the universe. Pages 152-154.)

The creation of an adequate wisdom comes from the observations of all forms and systems working together dynamically. Free will, in the form of ideas, regulations, policies and judgments, ought to reflect the interconnectivity of all frames of reference, thus providing the bases upon which to promote the flourishment of all living things, as well as planet Earth.

Ruminations on adequate wisdom

The basic tools of adequate wisdom simply attempt to cod-ify the major forms and processes in the universe, while pointing to the major ideas engendered by such forms and processes. But from this overview of form and process we discovered that there appears to be a possible equivalence between the two. The fixity of forms is often modified by the changing processes, allowing evolution to proceed, while the very nature of some structural arrangements seems to initiate process.

This peculiar arrangement between form and process sug-gests an apparent interdependence between steadfastness and change. One cannot have process without form, nor can we have form without process. There seems to be an ongoing relationship between fixity and fluidity.

Adequate wisdom views all forms and processes as integral to the understanding of existence. Physical forms and pro-cesses lead to biological forms and processes, which lead to human forms and processes, and which finally lead to social forms and processes.

Adequate wisdom points to the synergistic and hierarchical arrangements of forms and processes in all categories, be it chemical elements, plants and animals, human behaviors or social organizations.

Adequate wisdom discusses the phenomenon of recurring forms and processes that appear eternal or long lasting and which influence each other interdependently. Examples of these recurring forms and forces include: the universe itself and its fundamental forces; particles, atoms, mole-cules; galaxies, stars, solar systems and planetary motions and orbits; DNA master codes and evolution; atmosphere, geosphere and magnetosphere; laws, constitutions and corporate and religious regulations and edicts. All these recurring forms and processes affect human life and must be considered when formulating ideas and policies.

There is a tendency to disregard some of the physical recurring forms and processes when studying human experience. This is a mistake, which often leads to half-truths about existence. All forms and processes contribute to what we are today, what we were in the past and what we will be in the future. We are inexorably connected to all the variables of existence.

Because of the coalescence of forms, processes and ideas, we must acknowledge that human actions and behaviors are the results of the interplay among them.

Apart from scientific, medical and technological advances, how has the human being changed from his or her counterpart 5,000 years ago, about the beginning of recorded thought? From the writings and literature of the past, we gather that personality types were the same; that the goals of safety, shelter, protection and food production were the same; that the study of philosophy and religion existed as it does today. We can therefore deduce that the same recurring forms and processes occurring today were approximately the same 5,000 years ago (or longer). This is of course a nod to history partially repeating itself.

This recurrence suggests that the personal traits we experience today, both good and bad, will recur throughout the next 5,000 years and beyond (if our species is fortunate enough still to exist).

The changes that have occurred throughout history are the result of human design and free will, as well as the influence of chance and synergy; so that technology, fashion, art, music, fads and all other design-based ideas and creations do not remain as fixed recurring forms and processes. Because of this, new things in the world will always occur.

Because of recurring forms and processes, each individual human being is somewhat subjected to forces beyond his or her control, and because of this the idea of individual responsibility for one's actions (while a politically correct notion) must be modified. A large part of someone's actions is based upon personality, emotion, habit, character traits and social structures; and we have seen that many of these actions are essentially out of one's control. No infant, child, nor youth plans his or her personality; rather, it is given. A person who wears his heart on his sleeve, so to speak, will act out in a much different way than someone with a cool, detached personality. Therefore many actions are trait-driven and the individual should not be blamed for certain behaviors that most likely are not directed by volition and free will.

We have not spoken much about a person's character, which includes one's personality and temperament, but also includes years and years of experience facing the world and interacting in it. Character is manifested by the accumulation of deeds and concerns for other people and the environment. It can transcend personality by using will power to do good things or, for that matter, evil things.

Since we all have smatterings of various personality traits, some stronger, some weaker, it is the human character that truly defines each member of our population. According to the principles of adequate wisdom, a person of exceptional or good character recognizes the factors influencing the physical, biological, personal and social categories and does what he or she can to improve the lot of others, of living things, as well as the hydrosphere and atmosphere.

In today's very complex world, specialists abound everywhere and do not often examine seemingly disparate fields of study. There is an overabundance of the study of process, with form taking up the rear. Generalists are in short supply, but we need more of them to assess the interdependent assemblages of the physical, biological, personal and social categories through which forms and processes flow.

Adequate wisdom must address the nature of cosmic and biological evolution and the paradoxical movement toward heat death (positive entropy) versus the existence of synergistic wholes (negative entropy). The laws of thermodynamics and the arrow of time point to eventual loss of heat and the trend toward disorder. But, amazingly, the forms that abound in the universe feature negative entropy, drawing energy from other sources and allowing highly ordered forms, such as living things, to exist.

While certain particles, atoms and galaxies seem eternal, most forms eventually "die," including stars and living things. Thus, there exists a clock for negentropic forms and processes; yet while "alive" they defy the arrow of time.

What most scientists fail to comment on is the stunning synergistic nature of forms and processes. The holistic combination of forms and processes produces all of existence! Strings, quarks, quantum effects, the Higgs field/particle and the fundamental forces all combine to form new and greater wholes. *These events do not occur in isolation, but as an integrated whole.*

The complex nature of fusing hydrogen into helium and other elements in the core of stars represents a stellar synergy creating light, heat and other electromagnetic radiations, and finally planets and solar systems.

The synergistic combination of the Earth's core, mantle, crust, tectonic plates, the hydrosphere, the atmosphere, the Moon, Sun, planets and other non-stellar objects in the solar system all create and/or modify life on Earth.

The combination of RNA, DNA, organelles, cell replication and differentiation all form the supreme synergy called life.

As for predicting the future of existence on Earth, we have previously discussed the plight of the planet should a runaway greenhouse effect occur or should some other geological, biological or astronomical crisis occur. The future of the solar system, barring a nearby supernova, hinges on the ever growing size and heat from the Sun in a couple billion years and the eventual death of the Sun in the far future (billions of years ahead).

As for biological evolution, we have every reason to believe that what happened in the past, namely the extinction of whole species of life forms, could occur in the future. As humankind expands its destruction of the botanic world, more and more species will become extinct. If we escape from the impending climate crisis, there may be ages ahead of ice, ages where many species die off, and possibly, like the Cambrian explosion, a period producing an enormous number of new species. Thinking in millions of years, it seems quite possible that a superior new species will branch off from Homo sapiens, perhaps replacing it or, hopefully, cohabitating with it.

Thinking in a thousand or tens of thousands of years, the principle of synergy might continue to combine nations, races and ethnicities whereupon one day there may exist a holistic humanity, acting as if it were one collection of cells, inter-communicating and becoming a super-organism.

The proliferation of multinational corporations is a recent sign of synergy, wherein smaller companies are subsumed by the larger ones across all national and international borders. Add to this phenomenon the collectivities of nations forming communities or associations or pacts—synergy at work here again.

There is also the possibility of a secularization of all religions, a startling evolution which witnesses the diminution of icons, myths and rituals in favor of a unified belief in the eternal nature of existence and the flourishment of all species. Of course such a sea change that reshapes religions seems far-fetched, but may occur over the millennia.

As for developments in technology, one can only assume that there will be advanced computer-based intelligence and human-like robots (androids). Computer computations will soon exceed those in the brain, as astonishing as it may seem. We have already discussed the possibility of greatly expanded lifetimes thanks to advances in medical science, along with advances in space technology and habitation off the planet. And the amazing world of biotechnology, nanotechnology and genetic engineering offers humankind a dizzying array of possibilities for improved replacement organs, expanded lifetimes, as well as newly crafted life forms.

This writer believes that adequate wisdom will begin to make sense when we incorporate all the key forms, processes, ideas and beliefs into a unity which admittedly contains paradoxes, puzzles and counterintuitive observations. We must always be cognizant of the fact that all different forms and processes comingle in the lifetime of every single human being, group or civilization

Thus the goal of any idea system is the construction of policies that benefit all members of the human population, all living things, and the planet upon which we live.

More final thoughts and questions

In the 20th and 21st centuries, many ideas about the universe came from the cosmologists and physicists, painting a stunning, yet jumbled and confusing portrait of forces, particles, the spacetime continuum, parallel universes, waves, big bangs, relativity, cosmic inflations, quantum particles, quantum perturbations, quantum entanglement, vibrating strings and membranes.

There are astonishing theories that permit the creation of an infinite number of universes; that suggest that all matter and force may arise from infinitesimal strings and a mysterious Higgs field/particle that allow particles to exhibit mass.

From the study of quantum physics comes the incredible notion that the universe is somehow interconnected, no matter what the distance from one object to another may be. It seems possible that something informs the entire universe without the speed of light as a factor. This strange idea allows for the non-locality of time and space and thus the interconnectedness of the universe.

Evolutionary biologists show how humans and all other living things descended from previous living forms. They also describe the miraculous fertilized egg or germ cell unfolding into an incalculable number of living things. Many details of DNA-RNA and other cell forms and processes are fairly well known. New discoveries and techniques abound in the world of genetic engineering.

Other ideas about the universe and existence arrive from a variety of thinkers who point to the question of how life began, is there a God, is there meaning in the cosmos or only in the biosphere—specifically in the lives of humans.

If one were to ask how adequate wisdom might summarize existence, we would have to answer in a general manner;

at first claiming that it represents a study of forms and processes. It is that simple a concept. We look at the forms and processes throughout existence and begin to formulate ideas about them.

Can anyone fathom the incredible universe?

Does anyone really understand all existence?

Of course not.

We can create all kinds of ideas about the world, but we can never really know what is true. There are many questions we must continue to ask.

Can forms and processes allow for a belief in a transcendental existence that we cannot even begin to comprehend? Do stars, galaxies and clusters represent vast synergistic forms that take on an unknowable meaning?

Do strings, quarks, atoms and molecules arrange themselves in some fashion to signify interdependence upon each other so that we have form, structure and force in the world?

Is the entire universe, or our universe, capable of exhibiting action at a distance, as quantum entanglement suggests? Do all things really relate to all other things? Is there a reason for forms to coalesce into other, greater forms, as attested to by the synergistic trend of evolution? Is life merely a random event that can occur when an infinite number of universes are permitted? Is there anything to the concept of preformation? Do the initial conditions that existed at time of the big bang have built-in instructions to create specific forms and processes? Or, do random quantum fluctuations of particles set forth the conditions for stars, galaxies and solar systems (and perhaps other universes)?

If chance events alone create a universe, then from where do the forms arise? Do we have reliable ideas about the formation of strings, quarks and electrons? Since energy and matter are equivalent, we know that the energy unleashed by the Big Bang and inflation gave rise to particles. If infinity is viable, can we say with confidence that any manner of creation of particles and forces is possible?

In an eternal world, everything is possible. In a number of universes, it seems reasonable to assume that gravity and the other fundamental forces and forms will also exist to produce life forms. The question then is: are there any other universal fundamental forms and processes? Just because we cannot conceive of any, that does not rule them out.

But we are left with our own universe and the processes and structures therein. The great divide becomes visible when we pit quantum fluctuations and physical constants against a belief in a synergistic universe that continually becomes greater than its myriad number of parts.

We celebrate DNA as the blueprint for all life. But the DNA simply directs the establishment of various types of cells, which then direct the assemblage of tissue systems synergizing new forms called organs. These organs and organ systems go way beyond DNA and take on new responsibilities and functions. And when the final human form appears at birth there exists an organism that works thanks to its parts and yet takes on a distinctive form of its own. The newborn is flooded with stimuli from its environment and the universe into which it was born. As it matures, it exhibits character which makes it unique. The mature adult is a far cry from its DNA and organ systems. It has taken on its own synergistic form and flows interdependently into the river of other forms and processes. Could all these events be leading to an end state already built into the fabric of the universe?

Is the second law of thermodynamics going to ruin the fun? Will a heat death eventually occur to cause all stars to die? Or do all these conditions counterbalance one another? What about the production of the 92 elements from supernovas? Without the exceedingly hot furnaces within giant stars which eventually explode in stupendous fashion as they go through their death throes, there would not be life forms as we know it. In this very, very strange thing we call existence (including of course relativity and quantum physics), can any meaning be applied? Can any truth emerge at the universal level?

We know determinism and chance are big pieces of the puzzle. We also know that synergy—the formation of larger wholes and hierarchies—stands out as an overriding feature of existence. We also know that design exists in the biosphere (badgers creating a residence, bees forming hives, and humans designing cities).

Whether design exists in the physical world is up for grabs.

We also believe that free will exists in the human race, as the synergy between the body and the mind has created two powerful forces, called consciousness and cognition. Although some neurologists and biologists and physicists claim that humankind has no free will, the ability to compose a symphony, write a novel or build a museum cannot come from the random assembly of particles and neurons.

If infinity is real, then it seems possible that a universe among the infinite number of universes might contain forces that could conceivably control all the actions of its living forms, and thus free will would not exist. Of course, you might then ask, why not this one? That's a knotty problem, for sure, but I still put my money on human free will and design resulting from synergies within our central nervous system.

One must acknowledge the wonderful world of mathematics and the manner in which physical laws are described and actual events predicted. But with all the swirling forces and forms yielding to physical constants and determinism, there still remains a world of chance, design and free will. Are these latter attributes a result of the physical forces and forms? We would not be here if there were not a variety of physical forces and forms that helped to create us.

The determinism of DNA and the chance events throughout evolution helped finally to create a thinking being. Yes, we are all here because of the fundamental forces and forms and synergies. But, as a result of determined and designed phenomena, we are able to look at the parts and wholes and the various forces and events and are able to assemble them into ideas that help us understand existence.

Even though we were assembled by physical forces, we have flowered into remarkable living forms that resonate with and yet transcend all our antecedents.

One final thought, purely speculative, of course, is the notion that there may exist many fields or other phenomena beyond that which we already recognize that impart structure and process throughout all of existence. Even with our finely tuned scientific instruments, an ensemble of resonating influences may exist that we cannot observe because they cannot be discerned by the mind or deciphered by our instruments. For example, look at the strange neutrino, which emanates from the nuclear reactions in stars, and how the three types of this particle permeate space yet touch virtually nothing. Are they simply random events or do they contribute in some fashion to the structure of existence?

(Recent experiments in 2011 suggested that neutrinos travelled faster than the speed of light. If proven true, this would certainly raise eyebrows throughout the world of physics and cause a revision of some aspects of relativity theory.)

So we reach the obvious conclusion that we have much more to learn about the nature of existence, and this leads us to shy away from absolute truths while our brains keep scrutinizing the mysteries and marvels of the world.

Afterword

Because of the very nature of the subject matter, many essays in this work obviously require greater exposition and scrutiny. However, I hope I have offered a preliminary approach to the big picture and have pointed out some salient features, commonalities and questions about existence—all of which will hopefully provide the groundwork for a workable wisdom.

The obvious motif of this work is to embrace all components of the world and to suggest that everything is contingent upon everything else, while paradoxically allowing forms and processes to express themselves individually. The twelve variables of existence mentioned in the text could form the basis of a new idea system—call it adequate wisdom if you like—which allows us to see how everything is interrelated.

One important point to make is that we must always be wary of any ideas which claim to explain the world in absolute terms—including any such proclamations made in this work.

The information contained herein is based upon a lifetime of personal observations and thoughts, as well as books and other information sources. The author alone is responsible for errors, misstatements and other erroneous comments.

My thanks to the forms and processes that brought about my existence, as well as all other living and previously living people who have contributed to the rich and spectacular diversity we call existence. And to future members of our species, I welcome you and trust that you, too, will enquire about the nature of the world.

Glossary

Absolute values

These are values or beliefs which are transformative and which all members of the population ought to follow. Based upon common sense and the rational mind, all actions and policy making should incorporate goals that protect and nourish the biosphere, the hydrosphere and the atmosphere; improve the conditions of human beings; eliminate misinformation and falsehoods; balance pleasure and responsibility and promote strength and compassion.

Acorn effect

Just as an acorn contains all the information for building a magnificent oak tree, the "acorn effect" is meant to suggest that preformation of forms and processes is an idea to consider when establishing an adequate wisdom. In other words, could the chicken exist before the egg? On the grand scale, is there a plan for the unfolding of evolution?

Action at a distance

Quantum theory shows that there is a non-local characteristic of the universe, whereby action in one part of the universe can affect forms and processes in another part of the universe no matter how far away one event is from another. This concept has been shown to be valid by experimental observations. It also suggests that large forms, such as the solar system, can affect the components within it, including our biosphere. The theory also implies that the entire universe may be interconnected.

Adequate wisdom

This is the attempt to gather a workable set of guidelines based upon the rational mind and its observations of forms and processes which are all inexorably linked together. When all components of existence are unified, ideas arise that enable us to grasp an overview of existence and formulate policies beneficial to humanity. Obviously, adequate wisdom is a work in progress.

Aesthetics
The mind, through the operation of all five senses, appreciates and critiques art forms, music, literature, drama, fashion, the world of nature, and the beauty of all objects (including human beings and their appearance.)

Afterlife
There are those who believe in or hope for a life after death. Even though death ends the form and processes of a life form, an afterlife presupposes some unknown force or essence that moves on to a different realm of existence.

Agnosticism
The agnostic views the universe as a mysterious and indescribable phenomenon for which on the grand scale nothing is absolutely certain. Therefore, some unknown or unknowable entity or force *might* exist. Using adequate wisdom, an agnostic would *consider* the possibility of some ineffable essence that could be called God for want of a better word but would not necessarily believe that this God has any concern or affinity with humanity.

Allele
This represents two or more parts of a gene, each of which may signify a specific trait. Thus alleles can create dominant or recessive traits.

Altruism
Whether the care and concern for other people spring from a personality trait or some group of recurring forms and processes, altruism stands out as one of the finest achievements in human history. Helping other people is the height of compassion. Humanitarian efforts by individuals and groups soften the hard edges of life for many people, as well as other living things on the planet.

Amino acids

There are 20 key amino acid molecules that are coded by genes to produce the tremendous variety of proteins in the body, all of which have specific tasks to accomplish. Proteins essentially control all the processes in the cell. The shape of the proteins (in the form of globular structures) dictates their actions.

Amorality

When chance and/or determinism affect the forms and processes, there can be no moral judgments made. An earthquake or hurricane is amoral. An inherited disease is amoral. Perhaps personality and emotional traits are amoral, if we think that these have been given to us without our input. One can even consider a combination of moral and amoral actions when an individual using free will tries to modify certain given traits.

Adaptation

This a major evolutionary process that creates much of the diversity among living things, as life forms adapt to certain conditions over many generations, and even small changes can create new creatures.

Andromeda Galaxy

Also known as M31, Andromeda is the closest, largest spiral galaxy to the Sun, about 2.5 million light years away. This galaxy may contain up to one trillion stars. It is moving toward the Milky Way and may collide with our galaxy in about five billion years.

Animism

This is the belief that the physical universe and all living things are filled with a life force or essence which animates throughout the entire world. All things are to be treated as having a "soul" and as such deserve respect and/or worship.

Anti-Matter

Experiments in physics show that particles have an anti-matter component. For example, the negatively changed electron has an opposite positively charged positron. Theory suggests that the universe should have equal amounts of matter and anti-matter, and yet the great puzzle is why we do not observe anti-matter. When matter and anti-matter combine, they annihilate one another, so obviously it works out rather well that our part of the universe consists only of matter.

Atheism
One who espouses the complete disbelief in a God is the standard definition of an atheist. Since most of the human race is preoccupied with a belief in God, atheism is not a very popular belief system. However, it seems apparent that well educated people, especially those who believe only in rationality, do not often believe in a divinity, especially one that is responsible for all that exists in this strange universe. The problem with atheism is that it categorically denies the possibility of some unknown entity or force that may transcend our rational thinking processes. Adequate wisdom, while not waving the banner for a God, suggests that phenomena may exist beyond our comprehension and thus cannot be completely ruled out. Hence, agnosticism is the more sensible view as it admits to our incomplete understanding of the universe.

Atmosphere
The Earth's atmosphere consists of several layers that extend skyward for about 370 miles. The atmosphere absorbs energy from the Sun, protects us from harmful solar rays and provides a moderate climate for the planet. The troposphere is the first layer rising about 9 miles above the Earth, and in this region most weather patterns occur. It is separated from the next layer by the tropopause. Just above the tropopause is the stratosphere, rising about 31 miles. It is in this layer that we find the ozone layer. Both the troposphere and the stratosphere contain more than 99% of the air. Fifty three miles above lies the mesosphere, where frigid temperatures are found (up to minus 93 de-

grees Celsius.) But, as we go higher into the atmosphere (into the thermosphere), temperatures can reach over 1700 degrees Celsius! The major components of the atmosphere consist of 78% nitrogen and 21% oxygen.

Atom

The primal atom comprises one proton and one electron, and is called hydrogen. When a neutron combines with the proton and the one electron, we have the basic structure of the major forms in the universe. Ninety two natural atoms can be formed in the interior of large exploding stars. The protons and neutrons form ever increasing shells surrounded by the tiny electrons forming fuzzy orbitals around the atom's nucleus. Atoms are building blocks, like Legos, which form simple to extremely complex structures, called molecules, by electron bonding.

Axon

Found is most of the neuron cells in the central nervous system, the axon is a long nerve fiber that transmits signals to other neurons. It also is found in bundles (called white matter) that connect the two brain hemispheres.

Beginning and ending

Conventional thinking suggests that all things begin and end, yet on the vast cosmic scale, especially when infinite universes seem possible, there is the strong possibility that nothing began and nothing will end; that existence is one continuous phenomenon, which can be called eternity. While individual beings and entities will end, other forms and processes will continue to exist ad infinitum.

Beta-Endorphin

This neurotransmitter is released when the body is in pain and helps dull pain. It promotes feelings of well-being. It binds to opium and morphine receptors in the brain.

Big Bang

The start of our universe is believed to be the result of an incredible amount of energy emanating from a tiny amount of space or void and taking a few minutes to form the universe filled with radiation and particles eventually creating galaxies and stars. The Big Bang occurred about 13.7 billion years ago, and we know this because throughout all space there exists cosmic microwave background radiation whose temperature indicates when the beginning of the universe occurred. No one knows how or why the Big Bang occurred, and many believe that it could not result from nothing, so therefore an assumption can be made that it may not be unique; that it is merely one of many admittedly bombastic events which occur in an infinite universe.

Biosphere
All living and replicating forms on the planet comprise the biosphere, including simple cells and complex organisms. The first cells emerged on Earth about 3.8 billion years ago. There are now about nine million different species living on the planet.

Black hole
When a large star (much larger than the Sun) collapses, the powerful force of gravity eventually creates a supernova, leaving behind a remainder that continues to collapse upon itself until not even light can escape its clutches. Matter near the black hole is sucked inside, never to be seen again. Black holes are thought to exist in the centers of most, if not all, galaxies; but their nature is not well understood. (Evidently, some radiations can seep out from black holes.)

Causality
When a process (cause) creates a result (effect), we see determinism at work. The rotation of the Earth (process) brings about sunrise (effect). In a strict mechanistic view of existence, all effects are brought about by causes.

Celestial frames of reference
When we view the solar system and the universe from the Earth, we employ the geocentric frame, using the horizon as one of the Great Circles to calculate the location of ce-

lestial objects. The heliocentric frame places the Sun in the center for the point of reference. Each frame is a viable way to observe the universe.

Celestial influence
Not only do the Sun and Moon affect the tides and weather patterns, but because of the quantum notion of action at a distance, the entire solar system may have an influence upon life on our planet. This idea has been a taboo subject for several centuries and prevents research by a combination of the social sciences and celestial mechanics.

Celestial mechanics
Astronomers employ celestial mechanics to study the rotation, revolution and precession of bodies in the solar system; the orbits of the planets, moons and other bodies; and the location of stars and other celestial forms.

Cognition
The term is generally applied to human thought processes, like problem solving, language, cogitation and reflection; the application of memory and the senses; and a wide variety of mental abilities using free will and human design.

Collectivities
Any formation of two or more people, associations, nations, and any other collection of forms.

Common sense
The use of the mind as based upon experiences and observations that seem to work and are mostly shared by other members of the population. On a very warm day, one would generally not wear winter clothing. When driving an automobile, one would stop at a red light. One would not generally bite the hand that feeds it.

Copenhagen interpretation
This represents an agreement among some physicists that one must not look at quantum mechanics as an objective overview of reality, but rather one of probabilities and measurement. Waves and particles are a duality and both are

needed to explain many events. The nature of existence is probabilistic.

Cosmology
The large-scale study of the universe and cosmic evolution, including the Big Bang, inflation, dark energy, dark matter, galaxies and stars.

Creationism
The belief that the entire universe and life forms were created by a supreme being. Most creationists subscribe to the biblical account of evolution and think that the Earth was created in a very brief amount of time, essentially discounting the factual evidence of evolution, including the findings of geologists and paleontologists.

Dark energy
The only thing about dark energy that is known is that it creates the rapid expansion of the universe and represents about 70% of the universe. It is a mysterious force that occupies all space.

Dark matter
This is another mysterious entity or form that serves, along with gravity, to hold the galaxies together. It is not any kind of matter known to science. It has been calculated to represent less than 25% of the universe, but this figure might be smaller since recent studies have shown that there may be as many as three times more stars in the universe than originally calculated. If this is true, then the customary amount of ordinary matter may be more than the 5% calculated today.

Determinants
Any form, idea or force that influences another form, idea or process. Chance and determinism are determinants. Idea systems that influence others are determinants.

DNA
This is the hereditary molecule found in most living things. It is the super-molecule that directs the production of some

nine million species, based upon the synergy or combination of four chemical bases (adenine A, guanine G, cytosine C, and thymine T). In human DNA, there are about 3 billion of these bases, and about 99% of these are the same for each human being. The order of the bases, like letters in the alphabet, provide the information for building an organism. Some life forms, specifically the viruses, are assembled by RNA and not DNA.

Dogma
Any idea that claims it is invulnerable to criticism or change.

Dopamine
One of neurotransmitters in the brain, this substance has many roles, including an influence upon memory, mood, motivation, cognition, sleep and sexual gratification.

Ecliptic
The path of the yearly revolution of the Earth around the Sun.

Equifinality
The concept that if an end-state (final form) has been preformed or predetermined, the manner to reach it can be achieved by many different means. This concept might possibly explain why an individual's lot in life could be influenced by destiny while at the same time subject to twists and turns created by the person's free will, as well as external events.

Epiphenomena
Events occurring independently of the physical world. These epiphenomenal events are not related to the actual physical event, but occur as a side-effect. For example, living creatures may arise as a result of other events with no direct causal connection.

Eternal digital lifetime
When a member of our population dies, his or her lifetime and the achievements and memorable moments therein

can be captured for eternity thanks to the internet and all the burgeoning social networks. Photographs, videos, biographical material and comments from friends and relatives can be stored for as long as the internet survives, which one would expect as long as humans continue to survive as a species.

Eternity
Because of the real possibility that the world is infinite, without beginning and end, then spacetime, stars, galaxies and universes would continue ad infinitum. While individual forms, like living creatures, will unfortunately die, others take their place, ad infinitum.

Faith
In the religious sense, faith indicates that one believes in God and the promise for an eternal life. Faith also indicates a belief in someone or something that will provide comfort, goodness or trustworthiness.

Fertilization
The union of a sperm and egg cell which creates the zygote. It is the first step in the formation of a living creature.

Fixity
Those forms and processes (and ideas) which appear steadfast and unchangeable.

Fluidity
Those forms and processes (and ideas) that are open to change and variability.

Foreordination
Like preformation, this is the concept that suggests that wholes have been destined in advance to become what they are.

Gametes
The parental sex cells—egg and sperm.

Gene

A molecule that contains a unit of heredity. It contains a patch of DNA and RNA which codes for a specific protein. Genes are responsible for the passing along of some traits as well as the formation of cells and tissue or organ systems. Generally, each gene is represented by two forms (alleles), so that one allele might be dominant and the other recessive.

Genetic engineering

This is the insertion of foreign or synthetic genes into the DNA of an organism with the goal of producing changes in the cellular structure, often times to produce genetically modified products, such as insulin, human growth hormone or food products. Genetically modified mice are often used to study a wide variety of human ailments, such as cancers, heart disease and arthritis.

Genotype

All or part of the genetic composition of an individual.

Gluons

The particles that mediate the strong nuclear force which binds quarks and the nucleus of all atoms together.

God

This concept can be used as a semantic tool when describing the entire universe, including all the mysteries and unknown components of existence. In this sense, God can represent all the forms, forces and ideas in the world—whether or not we are aware of them. God can also represent the cause, if any, of the universe; as well as the divinity that may or may not direct all or some actions and behaviors of humanity. In an infinite universe, God may not have any relevance, since everything simply exists without the need for a designer.

Gravity

Gravity is the chief agent of attraction among cosmic entities, like stars and galaxies, and allows solar systems to exist and living things to occupy space on a rotating planet. Gravity and dark matter help to sustain the collectivities of galactic clusters and larger arrangement of galaxies and stars. Gravity is responsible for the collapse of very large stars that form supernovas which create the chemicals of life, as well as the strange phenomenon of black holes.

Half-Truths
Generally, this refers to beliefs that surround one idea or one field of study to the exclusion of all others. A whole truth would signify the incorporation of all twelve variables of existence as an interdependent whole.

Higgs boson/field
A theoretical particle and field that may give mass to particles.

Holism
The notion that evolution moves forward through the synergistic coalescence of forms. Wholes are found throughout the universe and the biosphere. Each whole appears to become part of a larger whole, and while retaining its individual functions, it also becomes part of a greater functioning entity.

Holistic morality
The development of a moral code that underscores the fact that people are not completely responsible for their actions, that forms and processes (such as emotions and traits) highly influence human behavior, and therefore suggests that we temper our anger and need for revenge by acknowledging the frailty of humanity. Such considerations would call for kinder treatment of criminals in a more humane manner of incarceration.

Homo erectus
A descendant of Homo habilis, this precursor to Homo sapiens had a larger brain and body size than Homo habilis.

Homo habilis
One of the oldest of the genus Homo, living about two million years ago and was an ancestor to Homo erectus. The human-like creature was short with a cranium about half the size of modern-day humans.

Homo provectus
A hypothetical genus that branches away from Homo sapiens and either supplants us or co-exists with us in the far future. We know that we are not the final end of thinking beings; evolution and synergy will someday produce a newer branch or version of our species.

Homo sapiens
The last member of the genus Homo (at least for the present), representing a life form that uses a spoken and written language, devises tools and is capable of abstract thought. In hundreds of thousands of years from the present, a new version of Homo sapiens (*Homo provectus*—advanced Homo sapiens) may evolve and either co-exist with or supplant the present human race.

Hormones
These chemicals are released by glands or cells in the body that direct changes or events in living things, such as metabolism, puberty, reproduction, immune system, growth, mood swings and hunger.

Humanistic civilization
This is an idealized goal for all nations and people to form a holistic civilization based upon the balancing of pleasure and responsibility and strength and compassion.

Hydrogen
The most abundant element in the universe, accounting for 75% of all matter. It results from the synergistic combination of a negatively charged electron and a positively charged proton. When one neutron is added, an isotope of hydrogen is formed, called deuterium. Large clouds of hydrogen are in part responsible for the formation of stars

and galaxies. Hydrogen and carbon form extremely important organic chains that are mostly responsible for the existence of living things.

Hydrosphere
This is the term for all the water on and below the Earth. The oceans represent over 70% of the Earth's surface. There may as much or more water below the Earth. Thanks to our fortuitous location from the Sun and thus the availability of water, life has arisen and continues to flourish.

Ideas
In adequate wisdom, ideas stand for beliefs, values and common sense based upon the interaction of forms and processes. Ideas can also represent metaphysical notions, such as God and all other thoughts that may exist independently of the physical universe.

Individualism
Each form or process exhibits its own individual function and essence. It can stand alone and act as a unique phenomenon. (It also becomes part of greater wholes, so a form or process exhibits duality—its own special function along with its role as part of a larger function.)

Ineffable
That which is unknowable, and thus beyond the reach of the rational mind.

Infinity
M-Theory and quantum mechanics have suggested that the world may actually be infinite, that there exists an infinite number of universes or that one endless universe exists without beginning or end. If this is true, then varieties of us would necessarily exist because in an infinite universe, all combinations of matter, energy and processes would be permitted. The concept of infinity is absolutely stunning. The difficult part of accepting this notion is the fact that we humans and all other living things reach an end and most likely do not become part of eternity.

Inflation of the cosmos

Occurring a fraction of a second after the Big Bang, the universe expanded 100-fold and created the enormous vastness of spacetime and eventually the galaxies and stars. (The details of the Big Bang and Inflation are way beyond the scope of this work, so anyone interested must perform extra research.)

Institutions

In the social realm of humanity, any coupling of a few people or an enormous number of people will constitute an institution. Small groups, like friends and families are sub-institutional; whereas large organizations, such as the military, religion, cities, states, nations and international structures, represent the nature of a social institution. These contain rules, regulations and other qualities that allow it to persist and flourish—and influence the lives of all people.

Intelligent design

This idea looks at the amazing intricacies of the universe and living things and postulates that something beyond cosmic and biological evolution and natural selection is responsible for all that exists. The implication of many ID advocates is that a God has created all that we see, usually the Christian God. Other than denying the validity of evolution and natural selection, intelligent design has not much to say, since whatever did design the universe, if indeed that is the case, this special entity is unknowable and incomprehensible.

Kelvin

This is a temperature scale whose lowest point is absolute zero degrees.

Light year

The distance that light travels in one year, which is roughly 5.88 trillion miles (or about 10 trillion kilometers).

M-Theory

An advanced version of five string theories, which allows them to be unified into one theory that proposes ten dimensions of space and one of time, hypothetically allowing for an endless number of universes.

Magellanic Clouds

The small and large Magellanic Clouds are actually two small galaxies that orbit the Milky Way and are part of our local group of galaxies. They are slowly drifting towards us and may collide with the Milky Way in several billion years.

Magnetosphere

The swirling molten metallic outer core of the Earth produces strong magnetic fields which interact with the solar winds outside the atmosphere and act as a shield against harmful cosmic rays. Without this magnetic shield, it is doubtful that any advanced life on Earth would have appeared.

Meiosis

This cell division is found in sexual reproduction, where the sperm and the egg cells (gametes) are formed. Chromosomes are recombined in these cells, producing a different genetic structure in each sex cell. The normal cell division process is called mitosis

Metaphysics

This is a very broad term covering many different philosophical concepts. One major branch, ontology, studies the nature of being, of existence. The other branch, epistemology, studies knowledge and how we know what we know. In adequate wisdom, something said to be metaphysical simply means that there may exist a phenomenon beyond the physical world and thus not comprehensible to our senses.

Microwave background radiation
This is the leftover radiation from the Big Bang that permeates all of space and has become so attenuated in strength that it radiates in microwave form at a frigid 2.7 degrees Kelvin.

Mitosis
This is the non-sexual division of cells in the body taking place continuously.

Morality
Any set of principles that suggest the right and wrong way to live and behave constitutes a moral or ethical system. Morals are relative, as they are issued in various periods of time, by various governments, religions, families and many other institutions. Even peer groups construct their own moral universe, whereby some individuals set their own ethical standards which oftentimes are at odds with the prevailing social standards. In adequate wisdom, the balancing of pleasure, responsibility, strength and compassion results in a set of ethics that guide people toward a life which cares for others, which contributes to the flourishment of the species and the planet, and which urges people to temper their anger and rage from the deeds of others who are not completely responsible for their actions.

Multiverse
Another way of stating that there exists an infinite number of universes.

Nanotechnology
A nanometer is one-billionth of a meter, which means that the technology involved here deals with atom- and molecule-sized materials that may someday revolutionize medicine, electronics and organic chemistry.

Natural selection
Certain new biological traits in a population may improve the success rate of certain individuals and thus bestow an advantage that becomes favored by evolution. These ad-

vantageous traits are then replicated and may often lead to the generation of a new species that fits a particular ecological niche better than its predecessor.

Neurotransmitters

These are chemical signals in the neuron that send signals to other neurons that produce neurochemicals. There are many different neurochemicals that provide different stimulations within the body, including raising the heart rate and blood pressure during stressful events; making you feel good; preventing anxiety; improving memory; elevating mood and emotion; and providing pleasure and reducing pain.

Norepinephrine

This neurotransmitter helps people gain focus and control, especially those who have attention deficit disorders. It acts as an antidepressant and an anti-psychotic.

Nothingness

What an interesting concept! Every ounce of rationality and common sense precludes the possibility of nothingness. Our universe most likely evolved from a singularity comprised of something no one can understand, but to think that everything arrived out of nothingness is too much of a jolt to cognition. The possibility that our particular universe may fade away from a heat death and become nothing is also difficult to swallow, but the belief in an infinite number of universes that neither began nor will never end pacifies our troubled minds and allows the concept of nothingness to become just that—nothing.

Novelty

Chance, free will, human design and synergy provide all the necessary ingredients for creativity and for all innovations. On the cosmic scale, chance and synergy are responsible for different types of stars, galaxies and other astronomical phenomena. In the biosphere, chance by way of natural selection and the replication of errors in DNA have produced some nine million different species. In the human world, free will, human design and synergy have produced varieties of cultures, governments and policies.

Nucleosynthesis
This is the processes of forming new atomic nuclei from protons and neutrons—thus the formation of several isotopes of hydrogen at the time of the Big Bang. This process also takes place in the evolution of stars, where atoms are manufactured.

Ontology
The branch of metaphysics which deals with the nature of existence and the relationships among forms and processes.

Organelles
All the sub-components of a cell that allow it to live and function. About a dozen major organelles form the synergistic whole of the cell, which is both individualistic and at the same time part of greater wholes, like tissues and organs.

Orion arm
Technically, it is called the Orion-Cygnus arm, a minor spiral arm of the Milky Way which contains a certain planet on which we live. The arm is about 10,000 light years in length.

Parallel universes
This notion is based upon the belief in an infinite number of universes, which may of course suggest that an endless number of universes might exist right next to us without our knowledge of them. The Many-Worlds theory, based upon quantum mechanics, suggests that events in the past, present and future can take any number of paths, each one real and each represented by a parallel universe.

Particles
These are sub-atomic, and comprise quarks, electrons, neutrons, protons, photons, neutrinos and an entire zoo

of other particles, many of which exist for very short times. For more information, do research on bosons and leptons.

Parts
Parts are really sub-wholes. Each whole is a part of a greater whole. Parts of a human being include the central nervous system (including the brain), the heart, liver and all the other parts that form the whole being. But each human is a part of a family, a peer group, a business and so on. Particle parts form atoms; atomic parts form molecules— and so on.

Phenotype
This differs from the genotype (which is the genetic makeup) in that the phenotype represents the outward appearance of an organism, including behavior traits and character. As a greater whole, the phenotype transcends the genetic factors and becomes more than merely a compilation of genes, cells and organs.

Polymer
A very large molecule that contains repeating chemical units. In biology, long chain polymers are an important ingredient in the genetic processes within cells.

Precession
Along with the Earth's daily spin and its yearly revolution around the Sun, a lesser known motion of the Earth is its wobble like a top which causes the axis to trace out a circle of 360 degrees about every 26,000 years.

Precession of the equinoxes
In astronomy, one form of locating celestial bodies is the use of the First Point of Aries, which occurs exactly at the spring equinox. The appellation of Aries was used in ancient astronomy, since the background for the spring equinox was the constellation Aries. Precession of the Earth has shifted the location, so the constellation Aries is no longer the backdrop for the First Point of Aries. Nonetheless, the spring equinox is still the First Point of Aries (measured in

hours instead of degrees), so that those who follow astrology can rest assured that the spring equinox remains the beginning of the first of 30 degrees of the sign of Aries, on or about March 20, with the next eleven 30-degrees signs that follow. Precession has rotated the 12 signs of the Zodiac (the background constellations), but the First Point of Aries will always occur at the beginning of the spring equinox. (Some believe that the rotation of the constellations caused by the Earth's wobble signals the advent of new ages, every 2100 years or so. The wobble will soon have the constellation Aquarius as its background, and thus all the hype about the Age of Aquarius many years ago.)

Preformation

This controversial concept suggests that all of existence or parts of existence may have been pre-packaged, so to speak, so that what we see unfolding as evolution is really the result of some kind of prior design. This notion of pre-programming has been attributed to God or to instructions built into the very fabric of spacetime. It also presents us with the incomprehensible view that the past, present and future are not what we think they are; thus the universe might be fooling us into believing what may not be true.

Proxima Centauri

A small star about 4.2 light years from Earth, which has the distinction of being our Sun's closest neighboring star.

Quantum

The smallest amount of any activity in a physical transaction. The quantum represents a discrete amount of energy; that is, there is no continuous or arbitrary multiple but rather a discrete multiple of a physical constant of an exceedingly small size.

Quantum entanglement

An extremely complex phenomenon that may suggest some kind of interconnectivity and non-locality of the entire universe.

Quantum mechanics

Developed in the 1920s, quantum mechanics deals with the startling observation that particles can be waves and waves can be particles, and that events do not have a certainty to them, but rather a probability—leading to the uncertainty principle. And for the first time in the study of the physical world, the role of the observer or experimenter has an important effect upon the results of any study. This branch of mechanics has lead to many new studies, especially string theory.

Quantum fluctuations

Slight changes in energy and time, especially at the beginning of our universe, may be responsible for the enormous expansion of the universe during the inflationary phase.

Recombination (Cosmology)

This is the early epoch in the history of the universe where charged electrons and protons began to form neutral hydrogen atoms.

Recombination (Biology)

Groups of genes are shuffled in the offspring of organisms. DNA and sometimes RNA molecules may be broken and then recombined by similar or dissimilar units.

Recurrence

This phenomenon indicates that certain forms, processes and ideas persist over long periods of time and have an important influence on existing forms, processes and ideas.

Red Giant Star

These stars are close to the end of their existence and have exhausted their supply of hydrogen. Our Sun will become a red giant in perhaps five billion years, where its radius will become hundreds of time larger than our current Sun, and the Earth (and its neighbors) will no longer exist

Reductionism

Generally, each whole is simply the sum of its parts. There is no effort to consider the synergistic value of wholes and their connection to other wholes. Causality is the key word, and free will is assumed not to exist.

Relativistic mechanics
This study comprises Einstein's special and general theories of relativity and are discussed briefly in the text.

Replication
In biology, this simply means that cells are copied exactly according to the DNA instructions.

RNA
It is theorized that RNA preceded the advent of DNA to form the first living cells, which eventually gathered other organelles including the DNA molecule responsible for populating the planet. RNA also acts as a messenger by copying the DNA instructions, eventually allowing for the production of proteins which accomplish most of the body's functions.

Runaway Greenhouse Effect
Temperatures rise, as they have been dramatically rising for the past 100 years, from a build-up of carbon dioxide in the atmosphere. If carbon dioxide continues to increase, the temperatures on the Earth continue to rise until a point is reached where the major ice caps melt and the oceans rise. If Earth were to become a greenhouse trapping all the heat inside the atmosphere, there is the possibility of a runaway greenhouse effect negatively affecting life on Earth. The planet Venus had a severe greenhouse effect where the oceans boiled off and clouds of sulfuric acid now dominate the atmosphere. It is unlikely that such a deadly effect would occur on Earth, but efforts should be taken immediately to reduce all activities that release harmful gases should such a Venus-like scenario occur.

Secularization of religion
This theoretical concept is based upon the wisdom of future humanity which removes the myths and falsehoods of current religions and replaces such beliefs with a rational, commonsensical system of belief; yet allowing for a belief

in a God which symbolizes all the mysteries and questions about existence. There is no reason why evolution (science) and eternity (God) cannot co-exist.

Serotonin
This neurotransmitter helps to keep us happy.

Social contract
Even though large social institutions have an enormous effect on and control of humanity, these large social forms are obligated to provide freedom and opportunities for its members to succeed and flourish. Members are subject to reasonable regulations to prevent a Wild West mentality, where survival of the fittest outranks the equality for all citizens. Nations and governments that tyrannize their members must be neutralized and eliminated.

Spacetime
In relativity theory, space and time combine to form a four-dimensional entity which describes the vast universe. Space had been thought to include three dimensions; but string theory, especially M-Theory, calls for the unusual collection of ten space dimensions and one time dimension. The extra seven dimensions of space are so tiny that they fold and compact and offer the possibility of an infinite number of universes.

Steroids
These are several types of organic compounds that include cholesterol, the sex hormones and anti-inflammatory substances, as well as compounds that regulate the body's metabolism.

Synapse
A neuron passes along electrical or chemical signals via the synapse, sometimes on the axon or sometimes on the dendrites. This process enables the release of neurotransmitters.

Synergy

The combination of wholes to form larger wholes. The evolutionary trend seems to depend upon the linking together of particles, which form atoms, which form molecules, which form stars and galaxies, which form solar systems, which form planets, and which form (in at least one situation), a biosphere, which then creates human beings in groups, institutions and world communities.

Synergy and individualism

Forms and processes operate in a dual fashion. Each whole stands out in its own specialized manner, performing an assortment of tasks and operations; yet these forms and processes also become part of a larger whole, with its own unique characteristics.

Tectonic plates

Portions of the Earth's crust and upper part of the mantle "float" and drift over billions of years. The continents rest upon these plates, which can collide and create new mountain ranges, as evidence by the collision of the Asia and India plates some 55 million years ago, creating the Himalaya mountain range. Ocean plates also exist and can collide, forming, for example, the Mariana Trench in the Pacific Ocean. (All these plates are called the Earth's lithosphere.)

Teleology

The belief that all forms and processes are part of a design built into existence with a purpose or final end.

Thermodynamics

This study of heat and energy has produced a 2nd Law which states that energetic states can only lose energy as time goes by. This is the concept of entropy which shows us that we cannot unscramble an egg once broken, that events are not reversible. However, negative entropic forms can and do exist; namely, stars, galaxies and living things by taking energy from external sources.

Transcendence
In common usage, this suggests that an idea or form can rise above its apparent meaning or function and obtain a higher, greater or mystical representation.

Volition
This is the act of making a decision or choice or using one's will power to accomplish a task or action. Volition is generally assumed to arise from the free will of each person, but we cannot underestimate the influences (both internal and external to the self) that may have a factor in the process of volition.

Weak force
Often called the weak interaction, this is one of the four fundamental forces of nature, which has now been unified with electromagnetism and called the electro-weak force. The weak force is responsible for the radioactive decay of particles and is mostly responsible for the fusion of hydrogen in stars.

White matter
This is the collection of axons in the brain, which appear white when staining brain tissue. The grey matter is the collection of dendrites throughout the brain.

Wholes
These represent any form, from particle, to atom, to star, to planet, to human being, to social groups. Wholes have their own identity and function, and yet become part of larger wholes with their own identity and function.

Wholism
Another way of spelling holism, this is the study all the wholes in the universe and their interactions.

Zygote
This is the primal cell, the one formed by the combination of sperm and egg, which begins the development of the embryo.

Bibliography

Abell, George. *Exploration of the Universe.* New York: Holt, Rinehart & Winston. 1969.

Albert, Ethel M., et al. *Great Traditions in Ethics.* New York: American Book Company. 1953.

Armstrong, Karen. *History of God.* New York: Alfred A. Knopf. 1994.

astronomy.swin.edu.au/cosmos/

Bally, John and Bo Reipurth. *The Birth of Stars and Planets.* Cambridge University Press: 2006.

Bauman, Zygmunt. *Postmodern Ethics.* Oxford UK: Blackwell. 1993.

Benson, Michael. *Far Out.* New York: Abrams. 2009.

Bergson, Henri. *Creative Evolution.* Westport CT: Greenwood Press. 1911.

Blum, Harold F. *Time's Arrow & Evolution, 3rd edition.* Princeton University Press. 1968.

Borenstein, Seth. *A Striking Idea: 2 Moons Became One.* Associated Press. August 4, 2011.

Boswell, John. *Christianity, Social Tolerance & Homosexuality.* University of Chicago Press. 1980.

Bullock, Alan and Oliver Stallybrass, eds. *Harper Dictionary of Modern Thought.* New York: Harper & Row. 1977.

Calle, Carlos I. *The Universe.* Amherst: Prometheus Books. 2009

Calow, Peter. *Biological Machines.* New York: Crane Russak. 1976.

Chaisson, Eric. *Astronomy Today. Vol.I The Solar System.* 5th edition. Upper Saddle River NJ: Prentice Hall. 2005

Chaisson, Eric and Steve McMillan. *Astronomy Today. Vol. II. Stars & Galaxies.* 4th edition. Upper Saddle River, NJ. Prentice Hall. 2002.

Chang, Kenneth. *Twinkle, Twinkle, Perhaps Three Times. New York Times.* December 2, 2010.

Cohen, Richard. *Chasing the Sun.* New York: Random House. 2010.

Comte, August. *General View of Positivism.* New York: Robert Speller & Sons. 1978.

Crick, Francis. *The Astonishing Hypothesis.* New York: Charles Scribner's Sons. 1994.

Croswell, Ken. *Alchemy of the Heavens.* New York: Anchor Books. 1995.

damtp.cam.ac.uk/research/gr/public/cos_home.html

Dawkins, Richard. *The Blind Watchmaker.* New York: W.W. Norton & Company. 1987.

--------------------. *The Greatest Show on Earth.* New York: Free Press. 2009.

Dennett, Daniel C. *Breaking the Spell.* New York: Penguin Group USA. 2006.

--------------------. *Darwin's Dangerous Idea.* New York: Simon & Schuster. 1995.

Dewey, Edward R. *Cycles.* New York: Hawthorn Books. 1971.

Duncan, Ronald and Miranda Weston-Smith, eds. *Encyclopaedia of Ignorance.* Oxford UK: Pergamon Press. 1977.

Durant, Will. *The Story of Philosophy.* New York: Simon Schuster. 1926.

Fayer, Michael D. *Absolutely Small.* New York: AMACON. 2010.

Feibleman, James K. *Understanding Oriental Philosophy.* New York: Horizon Press. 1976.

Ferreira, Pedro G. *The State of the Universe.* London: Weidenfeld & Nicolson. 2006

Ferris, Timothy. *The Whole Shebang.* New York: Simon & Schuster, 1997.

Flam, Faye. *CT Scans Offer Evolution Data. Philadelphia Inquirer.* May 20, 2011.

--------------. *Moon Rocks Tell a Vivid Story. Philadelphia Inquirer.* July 20, 2009.

Fraser, J.T. *The Genesis & Evolution of Time.* Amherst: University of Massachusetts Press. 1982.

Gasperini, Maurizio. *The Universe Before the Big Bang.* Berlin: Springer-Verlag. 2008.

Gingerich, Owen. *God's Universe.* Harvard University Press. 2006.

Gleick, James. *Chaos.* New York: Viking. 1987.

Gorman, James. *"Scientists in Britain Say That Laughter Releases Endorphins..." New York Times,* September 14, 2011.

Greene, Brian. *The Fabric of the Cosmos. Space, Time, & the Texture of Reality.* New York: Alfred A. Knopf, 2004

-----------------. *The Hidden Reality. Parallel Universes & the Deep Laws of the Cosmos.* New York: Alfred A. Knopf. 2011.

Greenfield, Susan A. *The Human Brain.* New York: Basic Books. 1997.

Gribbin, John. *In Search of the Multiverse.* Hoboken NJ: John Wiley & Sons. 2009.

Gubser, Steven S. *The Little Book of String Theory.* Princeton University Press. 2010.

Halle, Louis J. *Out of Chaos.* Boston: Houghton Mifflin. 1977.

Harris, Sam. *The Moral Landscape.* New York: Free Press. 2010

Hawking, Stephen. *A Brief History of Time.* New York: Bantam Books. 1988.

Hawking, Stephen and Leonard Mlodinow. *The Grand Design.* New York: Bantam Books. 2010.

----------------------. *The Universe in a Nutshell.* New York: Bantam Books. 2001.

Hitchens, Christopher. *God is Not Great.* New York: Hachette Book Group. 2007.

Hochman, Anndee. *Into the Science of Thriving. Philadelphia Inquirer.* July 23, 2011.

Huppert, Felicia A., et al. *The Science of Well-Being.* UK: Oxford University Press. 2007.

Jantsch, Erich. *Design for Evolution. Self-Organization in the Life of Human Systems.* New York: George Braziller. 1975.

Kaler, James B. *The Ever-Changing Sky. A Guide to the Celestial Sphere.* New York: Cambridge University Press. 1996.

Kauffman, Stuart. *At Home in the Universe. The Search for the Laws of Self-Organization & Complexity.* Oxford University Press. 1995.

Kean, Sam. *The Disappearing Spoon.* New York: Little Brown. 2010.

Klir, George J. ed., *Trends in General Systems Theory*. New York: Wiley. 1972.

Kung, Hans. *Does God Exist?* New York: Doubleday. 1978.

Kushner, Harold. *Who Needs God*. New York: Summit Books. 1989.

LeDoux, Joseph. *Synaptic Self*. New York: Macmillan. 2002.

Lewis, John S. *Physics & Chemistry of the Solar System, 2nd edition*. Burlington MA: Elsevier Academic Press. 2004.

Liddle, Andrew and Jon Loveday. *Oxford Companion to Cosmology*. Oxford University Press. 2008.

Little, Lewis E. *Theory of Elementary Waves*. Gainsville FL: New Classic Library. 2009.

Lovelock, James. *Gaia*. Oxford University Press. 1979.

May, Rollo. *Freedom & Destiny*. New York: W.W. Norton Company. 1981.

Mayr, Ernst. *Evolution & the Diversity of Life*. Cambridge MA: Harvard University Press. 1976.

Merleau-Ponty, Jacques and Bruno Morando. *Rebirth of Cosmology*. New York: Alfred A. Knopf. 1976.

Monod, Jacques. *Chance & Necessity*. New York: Vintage/Random House. 1972.

Murchie, Guy. *Music of the Sphere, Vol. II*. New York; Dover Publications. 1967.

Narlikar, Jayant and Geoffrey Burbidge. *Facts & Speculations in Cosmology*. Cambridge University Press. 2008.

nasa.gov/centers/goddard/home/index.html

Osborne, Roger. *Civilization*. New York: Pegasus Books. 2006

Ouspensky, P.D. *Tertium Organum.* Kessinger Publishing. 1921.

Overbye, Dennis. *Tiny Neutrinos May Have Broken Cosmic Speed Limit. New York Times,* September 23, 2011.

Parker, Steve. *Natural World.* New York: Dorling Kindersley. 1994.

Paxton, John and Sheila Fairchild. *Calendar of Creative Man.* New York: Facts on File. 1979.

Perelman, Lewis J. *The Global Mind.* New York: Mason Charter. 1976.

PhilosophyIdeas.com

plato.stanford.edu/contents.html

physics.nist.gov/cuu/Constants/

ptable.com/

Shapiro, Robert. *The Human Blueprint.* New York: St Martin's Press. 1991.

Smolin, Lee. *The Life of the Cosmos.* New York: Oxford University Press. 1997.

Smuts, Jan Christian. *Holism & Evolution.* London: Macmillan. 1926. (Reprinted 1973 by Greenwood Press.)

solstation.com/stars/sol-sum.htm

spof.gsfc.nasa.gov/Education/wms1.html

stars.astro.illinois.edu/celsph.html

Von Bertalanffy, Ludwig. *General System Theory.* New York: George Braziller. 1968.

--------------------. *Perspectives on General System Theory.* New York: George Braziller. 1975.

Wade, Nicholas. *New Glimpses of Life's Puzzling Origins. New York Times*. June 16, 2009.

Ward, Peter D. and Donald Brownlee. *Rare Earth*. New York: Copernicus. 2000.

Ward, Ritchie R. *The Living Clocks*. New York: Alfred A. Knopf. 1972.

Warren, Rick. *The Purpose Driven Life*. Grand Rapids MI: Zondervan. 2002.

Weinberg, Steven. *The First Three Minutes*. New York: Basic Books. 1977.

Whitehead, Alfred North. *Process & Reality*. New York: Free Press. 1929.

Wilson, Edward O. *Naturalist*. Washington DC: Island Press. 1994.

---------------------. *Sociobiology*. Cambridge MA: Harvard University Press. 1975.

Yau, Shing-Tung and Steve Nadis. *The Shape of Inner Space*. New York: Basic Books. 2010.

Zimmer, Carl. *How Many Species? New York Times*. August 30, 2011.

Acknowledgements

My thanks to Peter Glaze, Elementary Studio, John Patrick Hunt and Alan Gorman for their assistance in assembling this work.

About the author

Ronald P. Smolin

The author has worked in public relations, journalism, writing, publishing and dramatic arts throughout his lifetime. He assisted in public relations and journalism activities for the war on poverty programs in Newark, NJ, and the New York City Human Resources Administration. He started several small book distribution companies, including International Ideas, Inc; and later expanded to form a larger publishing and distribution company, called Trans-Atlantic Publications, and is its current president. He later created and is president of Coronet Books Inc., which imports and markets academic books from many countries around the world. He also edited the *Directory of Public High Technology Corporations* and was co-author of *The $25,000 Challenge*, a trivia quiz book. He also wrote and directed two theater plays which premiered in Philadelphia. He holds a bachelor's degree in journalism in the school of liberal arts, Pennsylvania State University